# ADVANCED WIRELESS LAN

Edited by **Song Guo**

**Environmental Impact of Biofuels**
Edited by Marco Aurélio dos Santos Bernardes

**Published by InTech**
Janeza Trdine 9, 51000 Rijeka, Croatia



**Publishing Process Manager** Petra Zobić

**Technical Editor** Teodora Smiljanic

**Cover Designer** InTech Design Team

**Image Copyright** Henry Nowick, 2010. Used under license from Shutterstock.com

Additional hard copies can be obtained from orders@intechopen.com

Environmental Impact of Biofuels, Edited by Marco Aurélio dos Santos Bernardes
p. cm.
ISBN 978-953-307-479-5

For EU product safety concerns:
IN TECH d.o.o., Prolaz Marije Krucifikse Kozulić 3, 51000 Rijeka, Croatia,
info@intechopen.com or visit our website at intechopen.com.

# Meet the editor

Dr Song Guo received the Ph.D. degree in computer science from the University of Ottawa, Canada in 2005. He is currently a Senior Associate Professor at School of Computer Science and Engineering, the University of Aizu, Japan. His research interests are mainly in the areas of protocol design and performance analysis for computer and telecommunication networks, presently focusing on network modeling, security analysis, cross-layer optimization, and performance evaluation of wireless ad hoc and sensor networks for reliable, energy-efficient, and cost effective communications.

Dr Guo is a senior member of IEEE and a member of ACM. He serves in many international journal editorial boards, including the prestigious IEEE Transactions on Parallel and Distributed Systems and Wireless Communications and Mobile Computing.

# Contents

# Preface

A wireless local area network (LAN) is a data transmission system designed to provide network access between computing devices by using radio waves rather than a cable infrastructure. Wireless LANs are designed to operate in a small area such as a building or office complex. The past two decades have witnessed starling advances in wireless LAN technologies that were stimulated by its increasing popularity in the home due to ease of installation, and in commercial complexes offering wireless access to their customers. This book presents some of the latest development status of wireless LAN and provides an opportunity for readers to explore the problems that arise in the rapidly developed technologies in wireless LAN.

This book consists of a number of self-contained chapters. Chapter 1 proposes various sum-product decoding methods for the punctured convolutional codes for the IEEE802.11n wireless LAN. It aims at providing high speed decoder by exploiting the higher degree parity check polynomial. The proposed sum-product decoding schemes achieve better performance than the conventional method with much reduced complexity. Chapter 2 theoretically analyzes the medium access control (MAC) layer throughput with distributed coordination function (DCF) protocol in the IEEE 802.11b/a/g/n-based wireless LANs under a fading channel model. Chapter 3 studies the stability region of the enhanced DCF (EDCF) MAC protocol under various mobility levels. Chapter 4 provides a survey of the main techniques introduced to improve quality-of-service (QoS) performance in wireless LANs. It represents the state of the art about current studies on how to preserve QoS in contention-based EDCA IEEE 802.11e networks under heavy loads. Chapter 5 proposes a framework for interaction between real and virtual users in hybrid shared space, in which a QoS adaptation mechanism is implemented for networks with bandwidth limitation. Finally, Chapter 6 proposes an indoor optical wireless LAN system using space-division-multiplexing (SDM) and wavelength-division-multiplexing (WDM) techniques. It presents the fabrication details of a dedicated complimentary-metal-oxide-semiconductor (CMOS) image sensor to realize a compact, high-speed, and intelligent optical wireless LAN.

In summary, the topics on physical layer, MAC layer, QoS and systems included in this book are expected to benefit both practitioners working in wireless LAN systems and researchers as well as graduate students with interest in this area. The editor is grateful to all authors for their contributions to the quality of this book. The assistance of reviewers for all chapters is also greatly appreciated. The University of Aizu provided an ideal working environment for the preparation of this book. The editor also appreciates the support of publishing process managers of InTech.

**Song Guo**
Senior Associate Professor, School of Computer Science and
Engineering, The University of Aizu,
Japan

# 1

# Sum-Product Decoding of Punctured Convolutional Code for Wireless LAN

Toshiyuki Shohon
*Kagawa National College of Technology*
*Japan*

## 1. Introduction

The next generation wireless Local Area Network (LAN) standard (IEEE802.11n) aims for high rate data transmission such as 100Mbps to 600Mbps. In order to implement that rate, high speed decoder for the convolutional code for the wireless LAN standard is necessary. From the viewpoint of high speed decoder, sum-product algorithm is an attractive decoding method, since decoding rule of sum-product algorithm is simple and sum-product algorithm is suit for parallel implementation. Furthermore, sum-product decoding is a soft-in soft-out decoding. The combined use of sum-product algorithm and another soft-in soft-out processing may provide good performance such as turbo equalization (Douillard et al., 1995; Laot et al., 2001). However, sum-product decoding for the convolutional code of the wireless LAN can not provide good performance. To improve the performance, the sum-product decoding method for the non-punctured convolutional code of the wireless LAN has been proposed (Shohon et al., 2009b; 2010). In the wireless LAN, however, punctured convolutional codes are also used. Therefore, this paper proposes sum-product decoding methods for the punctured convolutional codes of the wireless LAN.

A sum-product decoding method for convolutional codes has been introduced in (Kschischang et al., 2001). The sum-product algorithm uses a Wiberg-type graph that represents a code trellis with hidden variables as code states and visible variables as code bits. In this case, the Wiberg-type graph is equivalent to the code trellis and the sum-product algorithm becomes precisely identical to BCJR algorithm (Berrou, C. et al.;C; Kschischang et al., 2001). This method only gives interpretation of BCJR algorithm as sum-product algorithm. To avoid confusion, the method of (Kschischang et al., 2001) is referred to as BCJR. This paper deals with sum-product algorithm that uses a Tanner graph that represents a parity check matrix of the code. This sum-product algorithm is the same as that for Low-Density Parity-Check code (Gallager, 1963; MacKay, 1999). The sum-product decoding method for recursive systematic convolutional codes has been proposed in (Shohon et al., 2009a). In the wireless LAN, the non-systematic convolutional code is used. For the non-punctured convolutional code of the wireless LAN, the sum-product decoding method has been proposed in (Shohon et al., 2009b; 2010). In this paper, for punctured codes of the wireless LAN, sum-product decoding methods are proposed.

This paper is constructed as follows. In section 2, the convolutional codes used in the wireless LAN are explained. In section 3, the sum-product algorithm for convolutional codes is explained. In section 4, the sum-product decoding method for non-punctured convolutional code of the wireless LAN is explained and decoding performance of that method for punctured codes are shown. In section 5 and section 6, the sum-product decoding methods for punctured codes of the wireless LAN are proposed. In section 7, the decoding complexity is discussed.

## 2. Convolutional code for wireless LAN

### 2.1 Non-punctured code

The convolutional code for the wireless LAN is a non-systematic code with rate 1/2 (IEEE Std 802.11, 2007). Let a sequence of information bits be denoted by $x_0, x_1, \cdots, x_{N-1}$, a sequence of parity bits 1 be denoted by $p_{1,0}, p_{1,1}, \cdots, p_{1,N-1}$, and a sequence of parity bits 2 be denoted by $p_{2,0}, p_{2,1}, \cdots, p_{2,N-1}$. Polynomial representation for each sequence is as follows.

$$X(D) = x_0 + x_1 D + x_2 D^2 + \cdots + x_{N-1} D^{N-1} \tag{1}$$

$$P_1(D) = p_{1,0} + p_{1,1} D + p_{1,2} D^2 + \cdots + p_{1,N-1} D^{N-1} \tag{2}$$

$$P_2(D) = p_{2,0} + p_{2,1} D + p_{2,2} D^2 + \cdots + p_{2,N-1} D^{N-1} \tag{3}$$

Parity bit polynomials are given by

$$P_1(D) = G_1(D) X(D), \tag{4}$$

$$P_2(D) = G_2(D) X(D). \tag{5}$$

For the wireless LAN standard, $G_1(D)$ and $G_2(D)$ are given by

$$G_1(D) = 1 + D^2 + D^3 + D^5 + D^6, \tag{6}$$

$$G_2(D) = 1 + D + D^2 + D^3 + D^6. \tag{7}$$

Polynomials $X(D), P_1(D), P_2(D)$ are also represented by $X, P_1, P_2$ in this paper.

### 2.2 Punctured code

In this section, puncturing method for wireless LAN will be explained. Puncturing is a procedure for omitting some of the encoded bits in the transmitter. The effect from puncturing will reducing the number of transmitted bits and increasing the coding rate. Figure 1(a) to Fig.1(b) shows the puncturing pattern for coding rate, $r = 2/3, 3/4$.

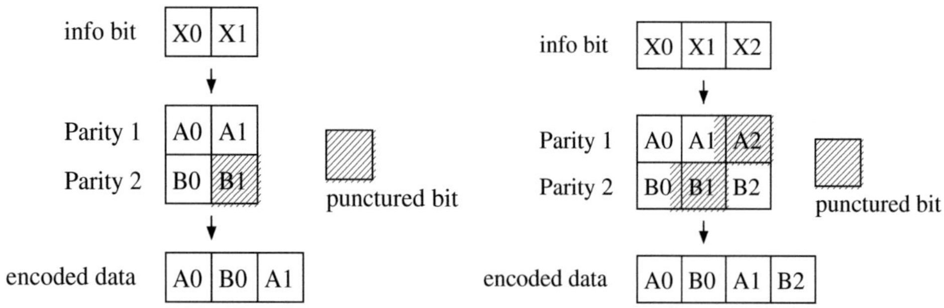


(a) Puncturing pattern for code rate 2/3 (b) Puncturing pattern for code rate 3/4

Fig. 1. Puncturing pattern

## 3. Sum-product algorithm for convolutional codes

Sum-product algorithm is a message exchanging algorithm along with edge of the Tanner graph of the code. Tanner graph is a bipartite graph that represents the parity check matrix of the code. For convolutional code, it is easy to obtain tanner graph from parity check polynomial. This section explains parity check polynomial for convolutional codes, tanner graph and sum-product algorithm.

### 3.1 Parity check polynomial of convolutional code for wireless LAN

From Equation 4 $\sim$ Equation 5, we can obtain following equations.

$$G_1(D)X + P_1 = 0 \tag{8}$$

$$G_2(D)X + P_2 = 0 \tag{9}$$

Let left parts of Equation 8 and Equation 9 be defined as parity check polynomial.

$$H_{org,1}(X, P_1) = G_1(D)X + P_1 \tag{10}$$

$$H_{org,2}(X, P_2) = G_2(D)X + P_2 \tag{11}$$

A tuple of polynomials $(X, P_1, P_2)$ is a code word if following equations are satisfied.

$$H_{org,1}(X, P_1) = 0 \tag{12}$$

$$H_{org,2}(X, P_2) = 0 \tag{13}$$

The degree of a parity check polynomial is denoted by $\nu$, that is the maximum degree of coefficients of the polynomial. For example, since coefficients of $H_{org,1}(X, P_1)$ are $\{G_1(D), 1\}$, the maximum degree is $\nu = 6$ that is the maximum degree of $G_1(D)$.

### 3.2 Tanner graph of convolutional code

From Equation 12, parity check equations at $k$ and $k+1$ time slots are given by

$$C_k \ : \ x_{k-6} + x_{k-5} + x_{k-3} + x_{k-2} + x_k + p_{1,k} = 0, \tag{14}$$

$$C_{k+1} \ : \ x_{k-5} + x_{k-4} + x_{k-2} + x_{k-1} + x_{k+1} + p_{1,k+1} = 0. \tag{15}$$

Those equations are corresponding to check nodes $C_k$ and $C_{k+1}$, of the tanner graph. The part of tanner graph corresponding to those parity check equations is as shown in Fig.2.

### 3.3 Algorithm

For convenience, bit node is denoted by $u_n$ such that

$$\begin{cases} u_{3n} = x_n \\ u_{3n+1} = p_{1,n} \\ u_{3n+2} = p_{2,n} \end{cases} \tag{16}$$

where information bit is $x_n$ and parity bits are $p_{1,n}$, $p_{2,n}$. Message from bit node, $u_n$, to check node $C_m$, is denoted by $V_{m,n}$. Message from check node, $C_m$, to bit node $u_n$, is denoted by $U_{m,n}$. Sum-Product algorithm is described as follows.

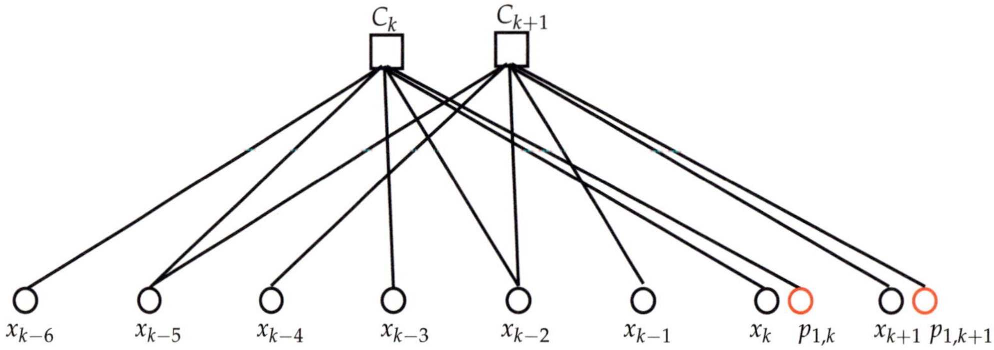


Fig. 2. Part of Tanner graph

Step1. Initialization

Each message $V_{m,n}$ is set to the initial value as follows.

$$V_{m,n} = \lambda_n = \frac{2r_n}{\sigma^2} \tag{17}$$

where, $r_n$ denotes received signal, $\sigma^2$ denotes variance of additive white Gaussian noise and $\lambda_n$ is channel value.

Step2. Message from check node to bit node

Each check node $C_m$ updates the message on bit node $u_n$ by gathering all incoming messages from other bit nodes that connected to check node $C_m$. Message $U_{m,n}$ is calculated by following equation (Gallager, 1963; Hagenauer, 1996; Richardson et al., 2001).

$$U_{m,n} = 2f_s \tanh^{-1}\left\{ \prod_{n' \in \mathcal{N}(m) \backslash n} \tanh\left(\frac{V_{m,n'}}{2}\right) \right\} \tag{18}$$

where, $\mathcal{N}(m)$ denotes the set of bit node numbers that connect to the check node $C_m$ and $f_s$ is a scaling factor. This factor is used in the proposed method described later. When $f_s$ is not specified, $f_s = 1$.

Step3. Message from bit node to check node

Each bit node $n$ propagates its message to all check nodes that connect to it.

$$V_{m,n} = \lambda_n + \sum_{m' \in \mathcal{M}(n) \backslash m} U_{m',n} \tag{19}$$

where $\mathcal{M}(n)$ denotes the set of check node numbers that connect to the bit node, $u_n$.

Step4. Tentative estimated code word computation

By summing up all the messages from all check nodes connected to a bit node, the a posteriori value $\Lambda_n$ can be obtained by

$$\Lambda_n = \lambda_n + \sum_{m \in \mathcal{M}(n)} U_{m,n}. \tag{20}$$

The extrinsic value, $L_e(u_n)$, of bit node $u_n$ can be obtained by

$$L_e(u_n) = \sum_{m \in \mathcal{M}(n)} U_{m,n}. \tag{21}$$

The tentative estimated bit $u'_n$ can be obtained by

$$u'_n = \begin{cases} 0 & if \quad \text{sign}(\Lambda_n) = +1 \\ 1 & if \quad \text{sign}(\Lambda_n) = -1 \end{cases} \tag{22}$$

Step5. Stop criterion

Tentative estimated code word $\mathbf{u}'$ obtained in Step 4 is checked against the parity check matrix $H$. If $H$ multiplied by Tentative estimated code word $\mathbf{u}'^T$ equal to zero vector, the decoder stop and outputs $\mathbf{u}'$, if not, it repeats Steps 2-5.

$$H\mathbf{u}'^T = \mathbf{0} \tag{23}$$

If maximum iteration number of decoding is set, the tentative estimated code word $\mathbf{u}'$ outputs after decoding procedure repeat the process until the maximum iteration is reached.

## 4. Sum-product decoding for wireless LAN (conventional method)

This section will give summary of (Shohon et al., 2009b; 2010). Sum-product decoding can be performed by using Equation 10 and Equation 11 as parity check polynomials. However, the decoding provides bad performance. Since the code under consideration is a non-systematic code, there are no received signals corresponding to information bits and channel values for information bits are zero. It can be seen from Equation 10, Equation 11 that each check node has more than one information bit connections. Therefore reliability increment at check node cannot be obtained. Consequently, conventional sum-product algorithm cannot realize good performance. To improve the sum-product decoding performance, I have proposed the 2-step decoding method (Shohon et al., 2009b; 2010).

### 4.1 2-Step decoding

The 2-step decoding method is as follows. (1) Only parity bits are decoded by sum-product algorithm. (2) With decoded parity bits, information bits are regenerated.

#### 4.1.1 Decoding parity bits

The parity check equation is derived from Equation 4 $\sim$ Equation 5 as follows.

$$G_2(D)P_1(D) + G_1(D)P_2(D) = 0 \tag{24}$$

The left part of the equation is defined as parity check polynomial $H(P_1, P_2)$.

$$H(P_1, P_2) = G_2(D)P_1 + G_1(D)P_2 \tag{25}$$

Parity bits $P_1$ and $P_2$ can be decoded by sum-product algorithm based on parity check polynomial given by Equation 25. By using the decoded parity bits, information bits can be regenerated.

#### 4.1.2 Decoding information bits

Decoded information bit $\hat{X}$ can be obtained by Equation 26 with decoded parity bits $\hat{P}_1, \hat{P}_2$.

$$\hat{X} = G_{x,1}(D)\hat{P}_1 + G_{x,2}(D)\hat{P}_2 \tag{26}$$

where,

$$G_{x,1}(D) = D^4 + D^2 \tag{27}$$
$$G_{x,2}(D) = D^4 + D^3 + D^2 + D + 1 \tag{28}$$

From Equation 26, Equation 27, and Equation 28, it can be seen that information bit can be regenerated by using a non-recursive convolutional encoder with input $\hat{P}_1$, $\hat{P}_2$ and output $\hat{X}$ as shown in Fig.3.

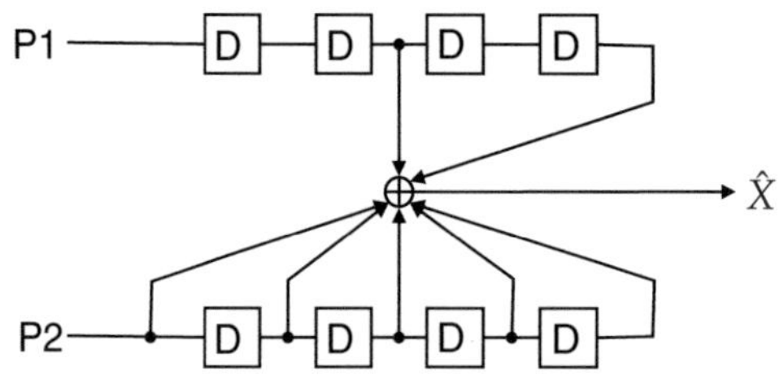


Fig. 3. Information bits regenerator

### 4.2 Higher degree parity check polynomial

I have proposed to use higher degree parity check polynomial to obtain further performance improvement (Shohon et al., 2009b; 2010).

The method is a sum-product decoding with higher degree parity check polynomial than that of the original parity check polynomial. In this section, the method is applied to improve the sum-product decoding performance for parity bits. The higher degree parity check polynomial is denoted by $H'(P_1, P_2)$, that is given by

$$H'(P_1, P_2) = M(D)H(P_1, P_2) \tag{29}$$
$$= M(D)G_2(D)P_1 + M(D)G_1(D)P_2 \tag{30}$$
$$= G'_2(D)P_1 + G'_1(D)P_2 \tag{31}$$

where $M(D)$ is a non-zero polynomial. Among possible higher degree parity check polynomials, we aim to select the optimum higher degree parity check polynomial by experiments and to use it for sum-product decoding. However, the number of prospective objects becomes too much when we include all possible higher degree parity check polynomials in the experimental objects. Therefore, we limit the range of degree of higher degree parity check polynomials ($\nu \leq 16$). For those higher degree parity check polynomials, we further limit the prospective objects by using $n_{fc}$, that is the number of four-cycles per one check node (Shohon et al., 2009a). For every degree of higher degree parity check polynomial, we select the higher degree parity check polynomial that has the minimum $n_{fc}$ among higher degree parity check polynomials of object degree and include it in the experimental objects. By this means, Table 1 was obtained.

| $\nu$ | $n_{fc}$ | $G_2'(oct)$ | $G_1'(oct)$ |
|---|---|---|---|
| 6 | 29 | 117 | 155 |
| 7 | 24 | 321 | 267 |
| 8 | 52 | 563 | 731 |
| 9 | 17 | 1067 | 1405 |
| 10 | 36 | 3131 | 2417 |
| 11 | 11 | 4015 | 6243 |
| 12 | 28 | 13103 | 16111 |
| 13 | 13 | 21003 | 30611 |
| 14 | 22 | 45203 | 65011 |
| 15 | 17 | 100001 | 145207 |
| 16 | 25 | 221001 | 322207 |

Table 1. Examined higher degree parity check polynomials for code rate 1/2

Experimental result shows that higher degree parity check polynomial of degree $\nu = 13$ provides the best performance. The higher degree parity check polynomial is given by

$$H'(P_1, P_2) = G_2'(D)P_1 + G_1'(D)P_2 \quad (32)$$

$$G_1'(D) = 1 + D^3 + D^7 + D^8 + D^{12} + D^{13} \quad (33)$$

$$G_2'(D) = 1 + D + D^9 + D^{13} \quad (34)$$

### 4.3 Simulation results for non-punctured code

Simulation condition is shown in Table 2. Hereafter, this condition was used, if simulation condition is not specified. Figure 4 shows simulation results. The figure shows that the performance for information bits of 2-Step Decoding with higher degree parity check polynomial (denoted by conventional) is only 0.7[dB] inferior to that of BCJR at bit error rate $10^{-5}$.

| | |
|---|---|
| Number of info bits per block | 1024[bit] |
| Termination | Zero-termination |
| Channel | Additive white Gaussian noise |
| Maximum iterations | 200 |

Table 2. Simulation condition

### 4.4 Simulation results for punctured codes

For non-punctured code, higher degree parity check polynomial with degree $\nu = 13$ provides the best performance. With that higher degree parity check polynomial, for punctured codes with code rates 2/3 and 3/4, the sum-product decoding simulation were executed. The simulation results are shown in Fig.5 and Fig.6.

From Fig.5 and Fig.6, it can be seen that the conventional method, that is sum-product decoding with higher degree parity check polynomial with $\nu = 13$, can not provide good performance for punctured code with code rates 2/3 and 3/4.

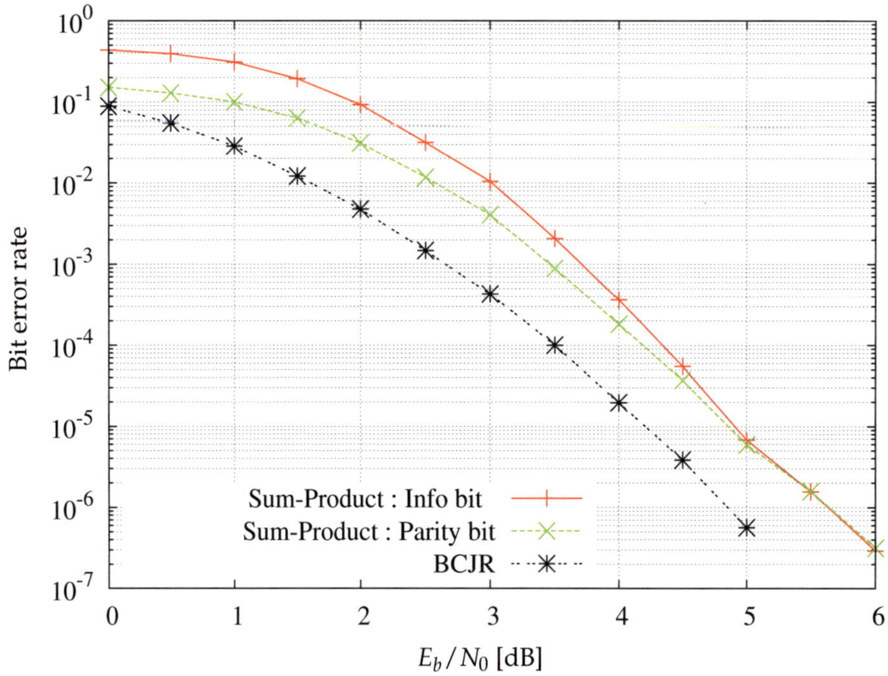


Fig. 4. Bit error rate performance of conventional method for code rate 1/2

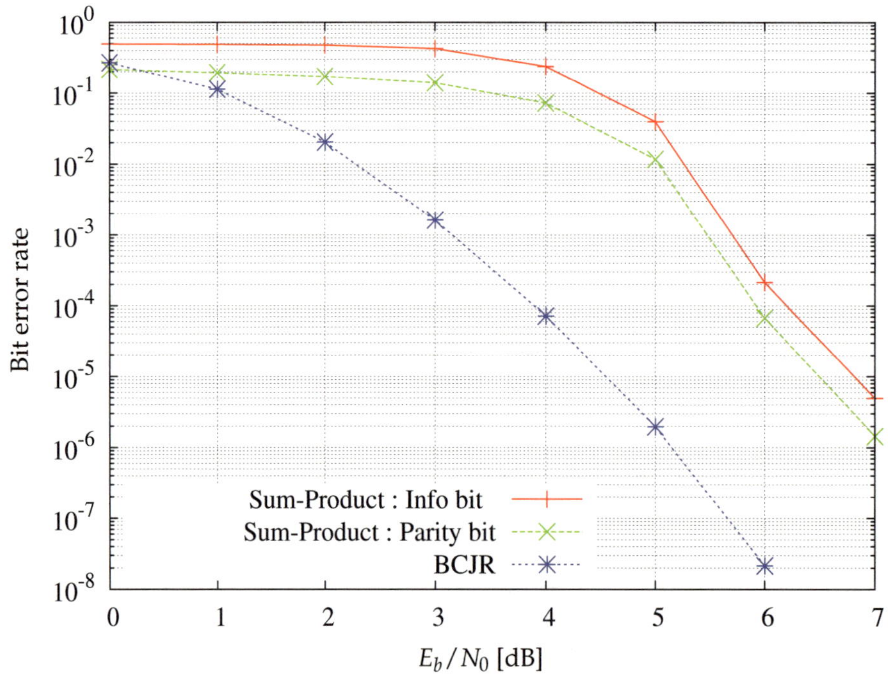


Fig. 5. Bit error rate performance of conventional method for code rate 2/3

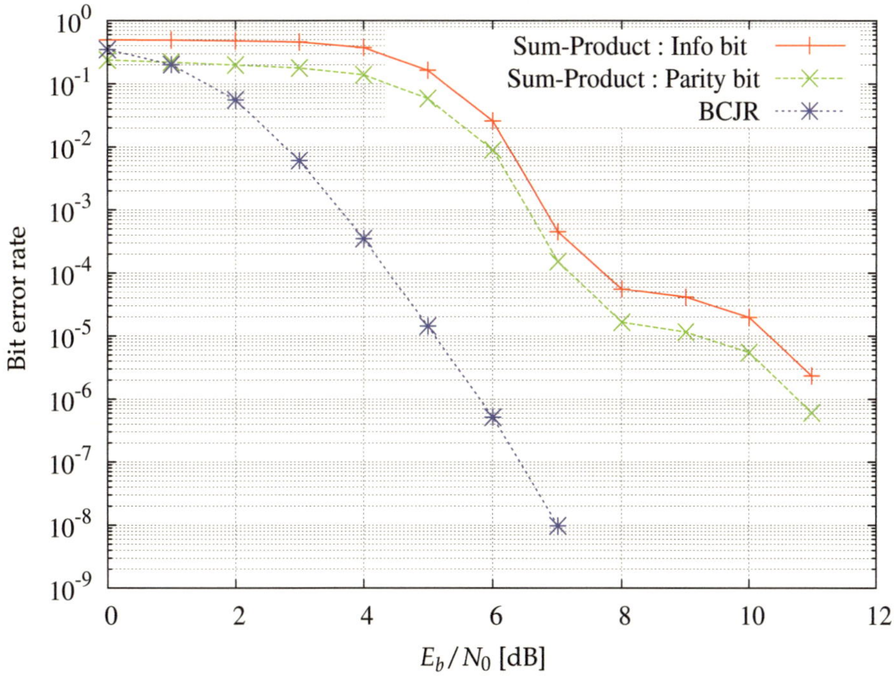


Fig. 6. Bit error rate performance of conventional method for code rate 3/4

## 5. Single punctured bit method (Proposed decoding method (1))

I inferred that the bad sum-product decoding performance for punctured codes is caused by more than one punctured bits included in the parity check equation at time slot $k$. The reason is as follows. Since received signals are not available for punctured bits, the channel values for punctured bits are zero. This causes that every messages from punctured bit node to check node are zero. In this case, like stopping set (Di et al., 2002), messages from the check node to bit nodes are zero. Therefore, sum-product algorithm does not work.

In order to improve the sum-product decoding performance, this paper proposes to use parity check equation that includes single punctured bit. The condition to include single punctured bit in parity check equation is referred to as single punctured bit condition. If single punctured bit is included in a parity check equation at time slot $k$, the message to the corresponding bit node can be obtained from the corresponding check node $C_k$. In this case, sum-product algorithm can work. Therefore, we expect that using higher degree parity check polynomial such that parity check equation includes single punctured bit, brings performance improvement of sum-product decoding of punctured codes.

### 5.1 Higher degree parity check polynomial satisfying single puncture bit condition

In this section, single punctured bit condition is derived for higher degree parity check polynomial. A higher degree parity check equation is given by

$$H'(P_1, P_2) = G'_2(D)P_1 + G'_1(D)P_2 \tag{35}$$

Generally, polynomials $G_1'(D)$ and $G_2'(D)$ are given by

$$G_1'(D) = \sum_{i=1}^{d_1} D^{\alpha_i} \quad (36)$$

$$G_2'(D) = \sum_{i=1}^{d_2} D^{\beta_i} \quad (37)$$

If $(P_1, P_2)$ is code word, it satisfies

$$G_2'(D)P_1 + G_1'(D)P_2 = 0 \quad (38)$$

From Equation 36, Equation 37 and Equation 38, parity check equation at time slot $k$ is represented by

$$\sum_{i=1}^{d_1} p_{1,k-\beta_i} + \sum_{i=1}^{d_2} p_{2,k-\alpha_i} = 0 \quad (39)$$

### 5.1.1 Code rate 2/3

For code rate $2/3$, punctured bits are

$$\{\, p_{2,2n+1} \mid n = 0, 1, 2, \cdots \} \quad (40)$$

From Equation 39 and Equation 40, it can be seen that punctured bits included in parity check equation at time slot $k$ satisfies

$$p_{2,k-\alpha_i} = p_{2,2n+1} \quad (41)$$

Therefore, we obtain

$$k - \alpha_i = 2n + 1 \quad (42)$$

$$\alpha_i = k - (2n + 1). \quad (43)$$

For time slot $k = 2l,\ l = 0, 1, 2, \cdots,$

$$\alpha_i = 2l - (2n + 1) \quad (44)$$

$$= 2(l - n) - 1 \quad (45)$$

Therefore, the set $\{\alpha_i \mid (\alpha_i \bmod 2) = 1\}$ in higher degree parity check polynomial corresponds to punctured bits in the parity check equation at time slot $k = 2l,\ l = 0, 1, 2, \cdots$. If Equation 46 is satisfied, the higher degree parity check polynomial satisfies single punctured bit condition at time slot $k = 2l,\ l = 0, 1, 2, \cdots$.

$$\#\{\alpha_i \mid (\alpha_i \bmod 2) = 1\} = 1 \quad (46)$$

where $\#\{x\}$ denotes the number of elements in the set $\{x\}$.

Similarly, if Equation 47 is satisfied, the higher degree parity check polynomial satisfies single punctured bit condition at time slot $k = 2l + 1,\ l = 0, 1, 2, \cdots$.

$$\#\{\alpha_i \mid (\alpha_i \bmod 2) = 0\} = 1 \quad (47)$$

Therefore, if either Equation 46 or Equation 47 is satisfied, the higher degree parity check polynomial satisfies single punctured bit condition.

### 5.1.2 Code rate 3/4

For code rate 3/4, punctured bits are

$$\begin{cases} p_{1,3n+2} \; n = 0,1,2,\cdots \\ p_{2,3n+1} \; n = 0,1,2,\cdots \end{cases}. \tag{48}$$

From Equation 39 and Equation 48, it can be seen that punctured bits included in parity check equation at time slot $k$ satisfies

$$\begin{cases} p_{1,k-\beta_i} = p_{1,3n+2} \; n = 0,1,2,\cdots \\ p_{2,k-\alpha_i} = p_{2,3n+1} \; n = 0,1,2,\cdots \end{cases}. \tag{49}$$

Therefore, we obtain

$$\begin{cases} k - \beta_i = 3n + 2 \\ k - \alpha_i = 3n + 1 \end{cases}. \tag{50}$$

For time slot $k = 3l,\ l = 0,1,2,\cdots,$

$$3l - \beta_i = 3n + 2 \tag{51}$$

$$\beta_i = 3(l - n) - 2 \tag{52}$$

$$3l - \alpha_i = 3n + 1 \tag{53}$$

$$\alpha_i = 3(l - n) - 1 \tag{54}$$

From Equation 52, it can be seen that the set $\{\beta_i \mid (\beta_i \bmod 3) = 1\}$ in higher degree parity check polynomial corresponds to punctured bits of parity bit $P_1$ in the parity check equation at time slot $k = 3l,\ l = 0,1,2,\cdots$. From Equation 54, it can be seen that the set $\{\alpha_i \mid (\alpha_i \bmod 3) = 2\}$ in higher degree parity check polynomial correspond to punctured bits of the parity bit $P_2$ in the parity check equation at time slot $k = 3l,\ l = 0,1,2,\cdots$.

Therefore, if either Equation 55 or Equation 56 is satisfied, the higher degree parity check polynomial satisfies single punctured bit condition at time slot $k = 3l,\ l = 0,1,2,\cdots$.

$$(\#\{\beta_i \mid (\beta_i \bmod 3) = 1\} = 1) \\ \wedge(\#\{\alpha_i \mid (\alpha_i \bmod 3) = 2\} = 0) \tag{55}$$

$$(\#\{\beta_i \mid (\beta_i \bmod 3) = 1\} = 0) \\ \wedge(\#\{\alpha_i \mid (\alpha_i \bmod 3) = 2\} = 1) \tag{56}$$

Similarly, if either Equation 57 or Equation 58 is satisfied, the higher degree parity check polynomial satisfies single punctured bit condition at time slot $k = 3l + 1,\ (l = 0,1,2,\cdots)$.

$$(\#\{\beta_i \mid (\beta_i \bmod 3) = 2\} = 1) \\ \wedge(\#\{\alpha_i \mid (\alpha_i \bmod 3) = 0\} = 0) \tag{57}$$

$$(\#\{\beta_i \mid (\beta_i \bmod 3) = 2\} = 0) \\ \wedge(\#\{\alpha_i \mid (\alpha_i \bmod 3) = 0\} = 1) \quad (58)$$

Similarly, if either Equation 59 or Equation 60 is satisfied, the higher degree parity check polynomial satisfies single punctured bit condition at time slot $k = 3l + 2$, $(l = 0, 1, 2, \cdots)$.

$$(\#\{\beta_i \mid (\beta_i \bmod 3) = 0\} = 1) \\ \wedge(\#\{\alpha_i \mid (\alpha_i \bmod 3) = 1\} = 0) \quad (59)$$

$$(\#\{\beta_i \mid (\beta_i \bmod 3) = 0\} = 0) \\ \wedge(\#\{\alpha_i \mid (\alpha_i \bmod 3) = 1\} = 1) \quad (60)$$

### 5.2 Search of higher degree parity check polynomial for decoding

In this paper, basically, the higher degree parity check polynomials for decoding are searched as follows.

Step.1 Select higher degree parity check polynomials with degree $\nu \leq 21$ that satisfies single punctured bit condition.

Step.2 Among those higher degree parity check polynomials, select the higher degree parity check polynomial that provides the best sum-product decoding performance by using computer simulation.

#### 5.2.1 Code rate 2/3

In the Step.1, 208 higher degree parity check polynomials satisfy single punctured bit condition. Since many higher degree parity check polynomials are selected, they are limited by $n_{fc}$. In this paper, among those higher degree parity check polynomials, 9 higher degree parity check polynomials with lower $n_{fc}$ are selected. They are shown in Table 3.

| No. | $\nu$ | $n_{fc}$ | $G'_2(oct)$ | $G'_1(oct)$ |
|---|---|---|---|---|
| 1 | 8 | 29 | 755 | 403 |
| 2 | 9 | 17 | 1067 | 1405 |
| 3 | 11 | 38 | 6143 | 5251 |
| 4 | 14 | 26 | 62501 | 50107 |
| 5 | 16 | 26 | 364203 | 202011 |
| 6 | 16 | 33 | 203133 | 310001 |
| 7 | 16 | 43 | 310207 | 243025 |
| 8 | 17 | 16 | 624403 | 500211 |
| 9 | 17 | 42 | 445207 | 640025 |

Table 3. Examined higher degree parity check polynomials for code rate 2/3

The simulation results of Step.2 with higher degree parity check polynomials in Table 3 are shown in Fig. 7. Simulation condition is shown in Table 2 and $E_b/N_0 = 5.0$ [dB]. From Fig. 7, it can be seen that higher degree parity check polynomial of No.5 with scaling factor $f_s = 0.9$ provides the best performance.

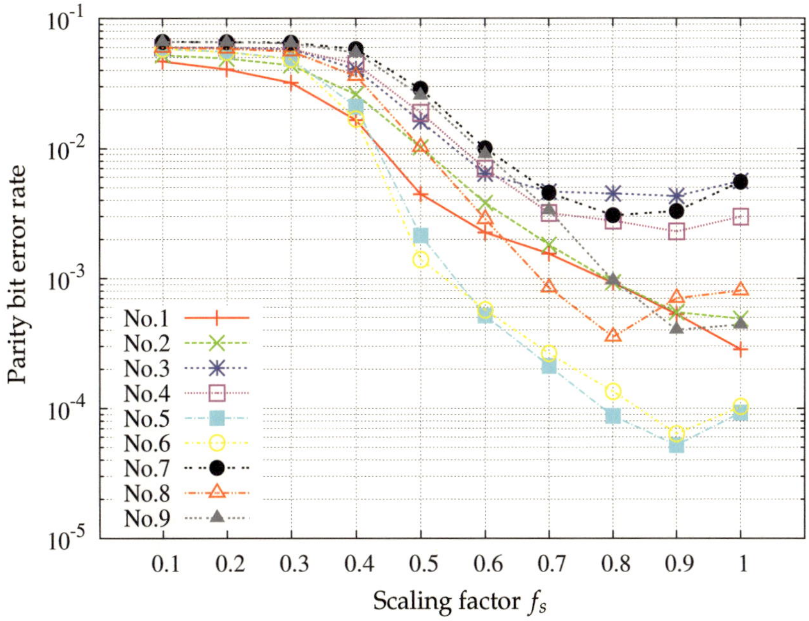


Fig. 7. Simulation results of Step.2 for code rate 2/3 at $E_b/N_0$ =5[dB]

### 5.2.2 Code rate 3/4

5.2.2.1 Step.1

For code rate 3/4, there are both puncture bits of parity $P_1$ and parity $P_2$. Decodable punctured parity bit by sum-product algorithm with certain parity check equation is either parity $P_1$ or parity bit $P_2$. From the viewpoint of decodable parity bit, single punctured bit condition can be arranged as follows.

1. If either Equation 55 or Equation 57, or Equation 59 is satisfied, the higher degree parity check polynomial includes single punctured bit of parity $P_1$. Therefore, with the higher degree parity check polynomial, punctured bits of parity $P_1$ can be decoded. That higher degree parity check polynomial is referred to as higher degree parity check polynomial for $P_1$.
2. If either Equation 56 or Equation 58 or Equation 60 is satisfied, the higher degree parity check polynomial includes single punctured bit of parity $P_2$. Therefore, with the higher degree parity check polynomial, punctured bits of parity $P_2$ can be decoded. That higher degree parity check polynomial is referred to as higher degree parity check polynomial for $P_2$.

Therefore, for code rate 3/4, both higher degree parity check polynomials for $P_1$ and $P_2$ are necessary to decode.

For code rate 3/4, there are 16 higher degree parity check polynomials for $P_1$ and 16 higher degree parity check polynomials for $P_2$. The number of combination of higher degree parity check polynomial for $P_1$ and that for $P_2$ is many. Therefore, they are limited by $n_{fc}$. Higher degree parity check polynomials that have lower $n_{fc}$ are selected as shown in Table 4 and Table 5.

| No. | $\nu$ | $n_{fc}$ | $G'_2(oct)$ | $G'_1(oct)$ |
|---|---|---|---|---|
| 1 | 12 | 30 | 14453 | 12121 |
| 2 | 21 | 64 | 17010055 | 10212103 |

Table 4. Examined higher degree parity check polynomials for $P_1$

| No. | $\nu$ | $n_{fc}$ | $G'_2(oct)$ | $G'_1(oct)$ |
|---|---|---|---|---|
| 3 | 7 | 24 | 321 | 267 |
| 4 | 12 | 29 | 11055 | 15103 |
| 5 | 21 | 38 | 10540055 | 14222103 |

Table 5. Examined higher degree parity check polynomials for $P_2$

5.2.2.2 Step.2

A block diagram of the decoder for code rate 3/4 is shown in Fig. 8. It is similar to a turbo decoder. In Fig.8, DEC1 is sum-product algorithm decoder with higher degree parity check polynomial for $P_1$ and DEC2 is sum-product algorithm decoder with higher degree parity check polynomial for $P_2$. Channel value is denoted by $\lambda_n$. Extrinsic values of DEC1 and DEC2 are denoted by $L_{e1}(u_n)$ and $L_{e2}(u_n)$, respectively. A posteriori value of DEC2 is denoted by $\Lambda_{2,n}$.

In DEC1, $L_{e2}(u_n)$ is added to $\lambda_n$ as follows.

$$\lambda'_n = \lambda_n + L_{e2}(u_n) \tag{61}$$

The value $\lambda'_n$ is used as initial value of $\lambda_n$ in Equation 17.

Similarly, in DEC2, $L_{e1}(u_n)$ is added to $\lambda_n$ and that value is used as initial value of $\lambda_n$. In computer simulation, the number of iteration of sum-product algorithm at each decoder was set to 1. The maximum number of iteration between two decoders was set to 200. Other simulation conditions are the same as shown in Table 2.

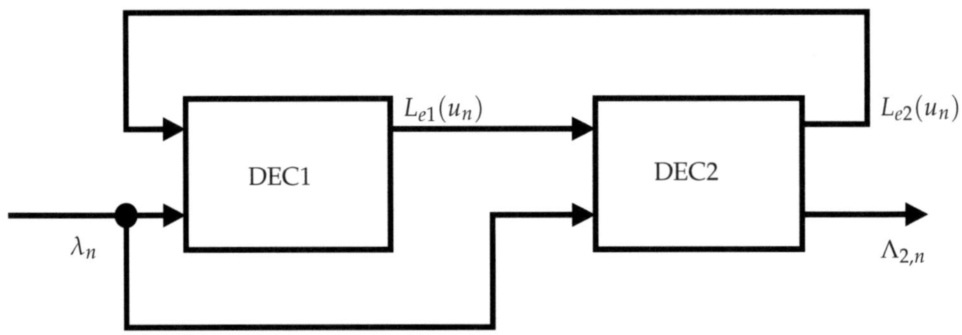


Fig. 8. Block diagram of decoder of single punctured bit method for code rate 3/4

Figure 9 shows the simulation results of Step.2 for code rate 3/4 at $E_b/N_0 = 6$[dB].

From Fig.9, it can be seen that the combination of higher degree parity check polynomials No.2 and No.3 with scaling factor $f_s = 0.5$ provides the best decoding performance.

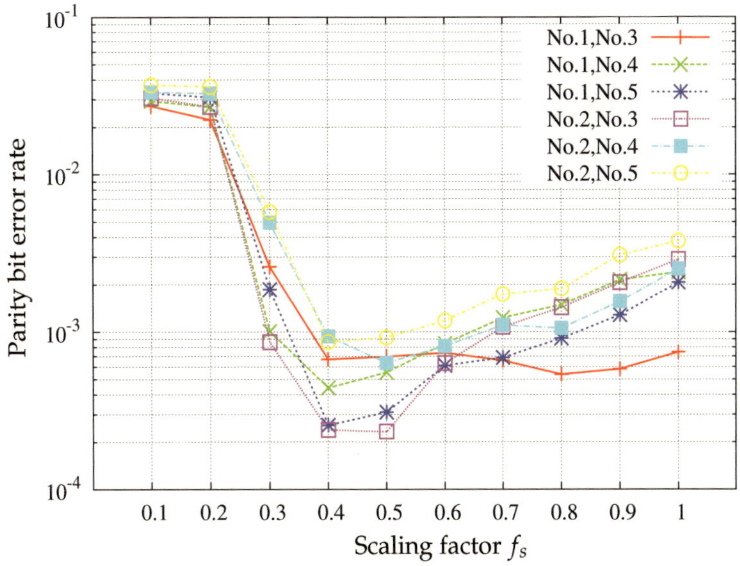


Fig. 9. Simulation results of Step.2 for code rate 3/4 at $E_b/N_0 = 6$[dB]

### 5.3 Simulation results

#### 5.3.1 Code rate 2/3

Figure 10 shows bit error rate performance of the single punctured bit method for code rate 2/3.

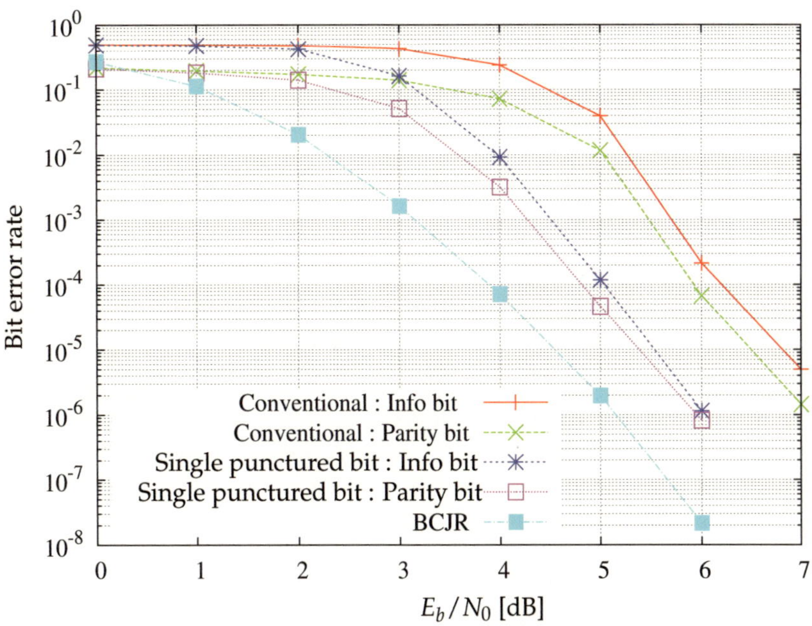


Fig. 10. BER performance of single punctured bit method for code rate 2/3

From Fig.10 it can be seen that the parity bit error rate performance of the single punctured bit method is 1.12[dB] superior to that of the conventional method (higher degree parity check polynomial of $\nu = 13$) at bit error rate $10^{-5}$. The parity bit error rate performance of the single punctured bit method is only 0.83[dB] inferior to that of BCJR.

From Fig.10, information bit error performance of the single punctured bit method is 1.28[dB] superior to that of the conventional method at bit error rate $10^{-5}$. The information bit error rate performance of the single punctured bit method is only 0.98 [dB] inferior to that of BCJR.

### 5.3.2 Code rate 3/4

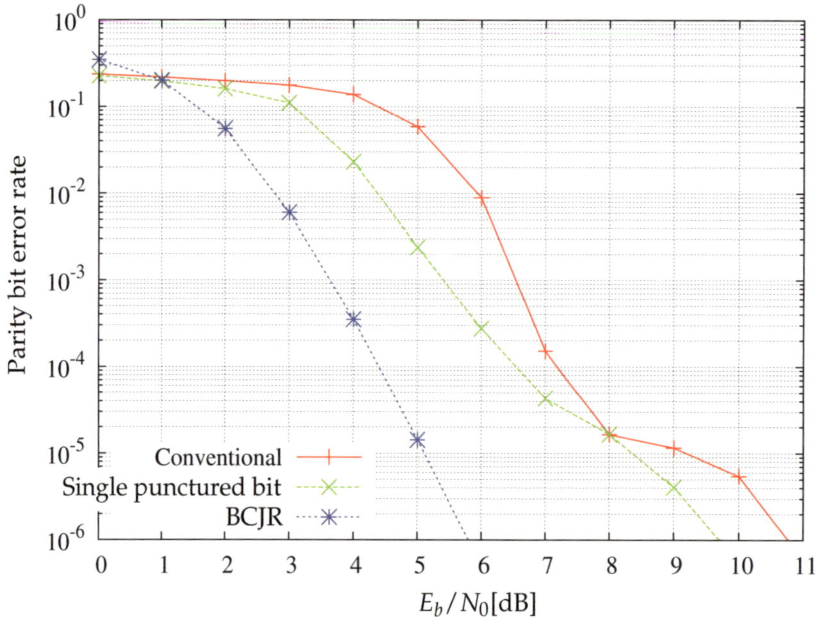


Fig. 11. Parity bit error rate performance of single punctured bit method for code rate 3/4

Figure 11 shows parity bit error rate performance of the single punctured bit method for code rate 3/4. From Fig.11 it can be seen that the parity bit error rate performance of the single punctured bit method is 0.82[dB] superior to that of the conventional method (higher degree parity check polynomial of $\nu = 13$) at bit error rate $10^{-5}$. The parity bit error rate performance of the single punctured bit method is 3.24[dB] inferior to that of BCJR.

Figure 12 shows information bit error rate performance of the single punctured bit method. From Fig.12, it can be seen that the information bit error rate performance of the single punctured bit method is 1.11[dB] superior to the conventional method at bit error rate $10^{-5}$. The information bit error rate performance of the single punctured bit method is 4.11[dB] inferior to that of BCJR at bit error rate $10^{-5}$.

## 6. Switching parity check method (proposed decoding method (2))

For code rate 3/4, the proposed method (1) can not provide good performance. Therefore, this paper try to improve the sum-product decoding performance for code rate 3/4.

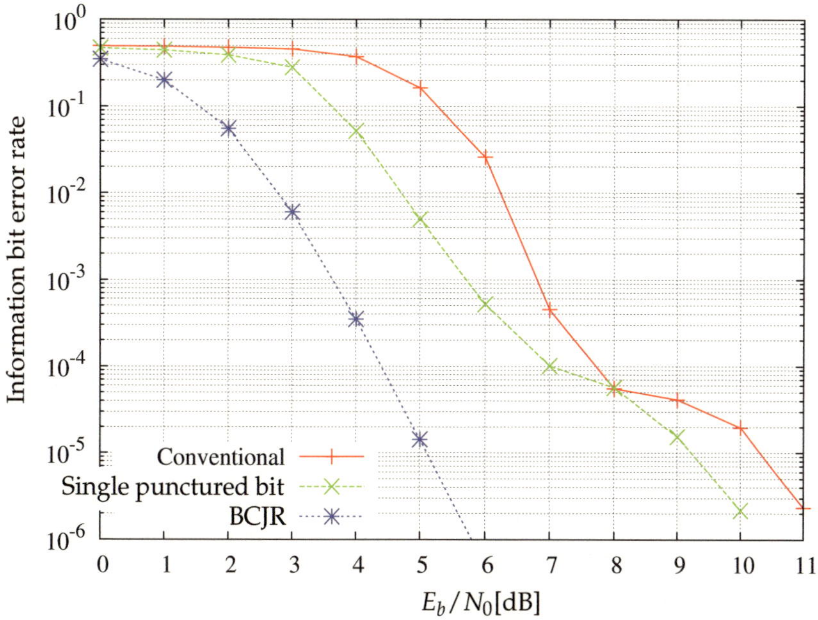


Fig. 12. Information bit error rate performance of single punctured bit method for code rate 3/4

I inferred that the bad decoding performance is caused by the four-cycles of higher degree parity check polynomial, since $n_{fc}$ of higher degree parity check polynomial satisfying single punctured bit condition tends to be larger than $n_{fc}$ of higher degree parity check polynomial that does not satisfy single punctured bit condition. Therefore, this paper proposes following method. Only at first iteration, the higher degree parity check polynomial satisfying single punctured bit condition is used to decode and after first iteration, another higher degree parity check polynomial without single punctured bit condition is used to decode. By decoding, only at first iteration, with higher degree parity check polynomial satisfying single punctured bit condition, the a posteriori values of punctured bits are obtained. After obtaining the a posteriori values of punctured bit, the higher degree parity check polynomial with lower $n_{fc}$ may provide good bit error rate performance.

Figure 13 shows a block diagram of decoder of the switching parity check method. In Fig. 13, DEC1 is a sum-product algorithm decoder with higher degree parity check polynomial for $P_1$, DEC2 is a sum-product algorithm decoder with higher degree parity check polynomial for $P_2$ and DEC3 is a sum-product algorithm decoder with higher degree parity check polynomial with lower $n_{fc}$ for iteration. Chanel values for DEC1, DEC2 and DEC3 are $\lambda_{1,n}$, $\lambda_{2,n}$ and $\lambda_{3,n}$, respectively. A posteriori values of DEC1, DEC2 and DEC3 are $\Lambda_{1,n}$, $\Lambda_{2,n}$ and $\Lambda_{3,n}$, respectively. Decoders DEC2 and DEC3 use the a posteriori value of previous decoder as the channel value.

### 6.1 Search of higher degree parity check polynomial for decoding

This paper searches higher degree parity check polynomials for DEC1, DEC2 and DEC3 by computer simulation.

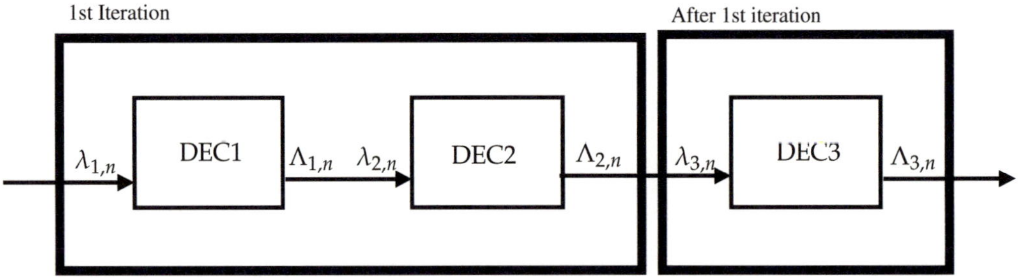


Fig. 13. Block diagram of decoder of switching parity check method for code rate 3/4

In this paper, the higher degree parity check polynomials for DEC1, DEC2 and DEC3 were selected from Table 4, Table 5 and Table 1, respectively. There are many number of combination of the higher degree parity check polynomials. Therefore, this paper searches the higher degree parity check polynomials as follows.

Step.1 At first, the higher degree parity check polynomial for DEC3 is determined by decoding simulation with only DEC3.

Step.2 With the determined higher degree parity check polynomial for DEC3, the higher degree parity check polynomials for DEC1 and DEC2 are determined by decoding simulation with DEC1, DEC2 and DEC3.

Figure 14 shows the simulation results of Step.1 at $E_b/N_0 = 6$[dB]. From Fig.14, it can be seen that the higher degree parity check polynomial with $\nu = 15$ provides the best performance. Therefore, that higher degree parity check polynomial is used.

Figure 15 shows the simulation results of Step.2 at $E_b/N_0$=7[dB]. From Fig.15, it can be seen that the combination of higher degree parity check polynomials No.2 and No.5 with scaling factor $f_s = 0.1$ provides the best performance, where scaling factor $f_s = 0.1$ is used for DEC1 and DEC2, and DEC3 uses fixed scaling factor $f_s = 1$.

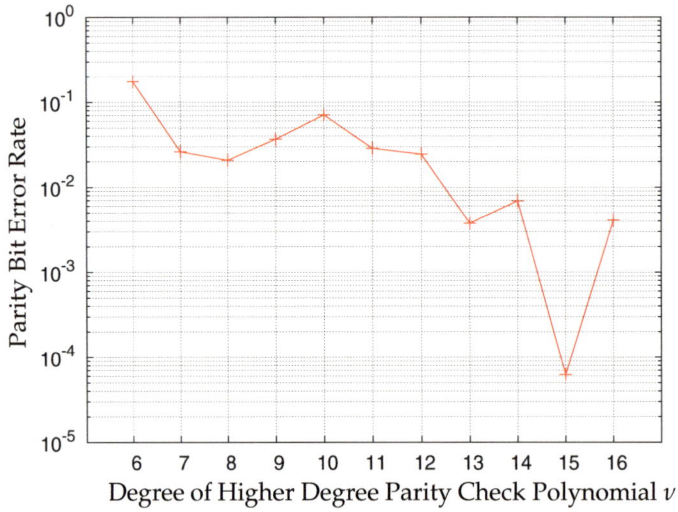


Fig. 14. Simulation results of step.1 in switching parity check method at $E_b/N_0 = 6$[dB]

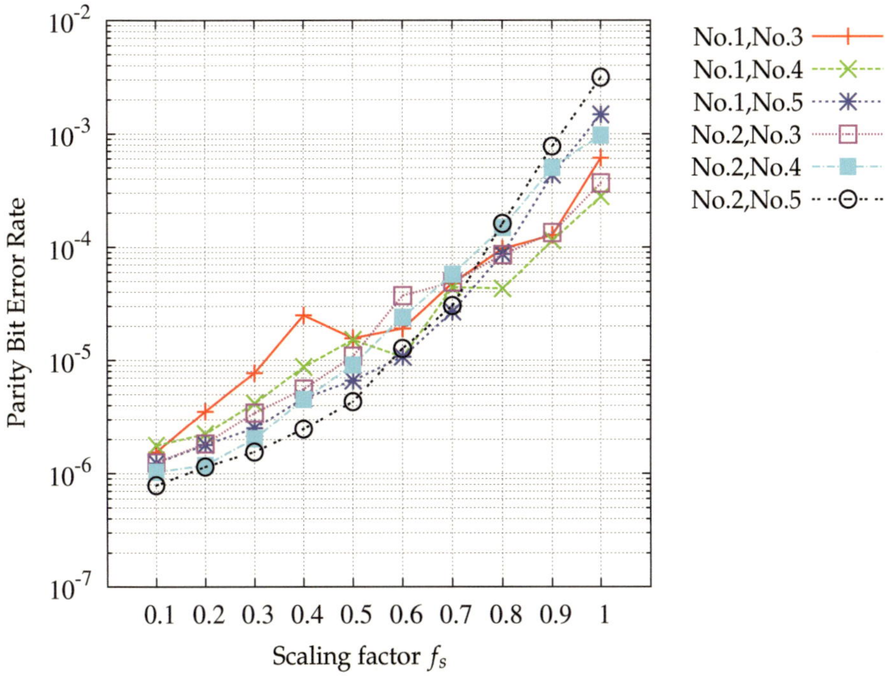


Fig. 15. Simulation results of step.2 in switching parity check method at $E_b/N_0 = 7$[dB]

### 6.2 Simulation results

Simulation results are shown in Fig.16 and 17. Figure 16 shows parity bit error rate performance. From Fig.16, it can be seen that parity bit error rate performance of the switching parity check method is 3.02[dB] superior to that of the conventional method and 2.2[dB] superior to that of the single punctured bit method. Parity bit error rate performance of the switching parity check method is only 1.04[dB] inferior to that of BCJR.

Figure 17 shows information bit error rate performance. From Fig.17, it can be seen that information bit error rate performance of the switching parity check method is 4.16[dB] superior to that of the conventional method and 3.05[dB] superior to that of the single punctured bit method. Information bit error rate performance of the switching parity check method is only 1.06 [dB] inferior to that of BCJR.

## 7. Decoding complexity

Table 6 and Table 7 show the numbers of operations per one bit decoding for sum-product algorithm and BCJR, respectively. In both tables, $N_{add}$ denotes the number of additions, $N_{mult}$ denotes the number of multiplications and $N_{total}$ denotes the total number of operations. For sum-product algorithm, $N_{sp}$ denotes the number of operations for $\tanh(\cdot)$, $\tanh^{-1}(\cdot)$. For BCJR, $N_{sp}$ denotes the number of operations for $\exp(\cdot)$, $\log(\cdot)$. In Table 6, for information bits, $N_{add}$ shows the number of XOR's. In Table 6, $N_{itr}$ denotes the average number of iterations, where the number was counted at $E_b/N_0$=6[dB] by using computer simulation. For code rate 2/3, complexity of the single punctured bit method is shown. For code rate 3/4, complexity of the switching parity check method is shown. It is necessary to notice that iteration of sum-product algorithm is required for parity bits decoding only. For the switching parity

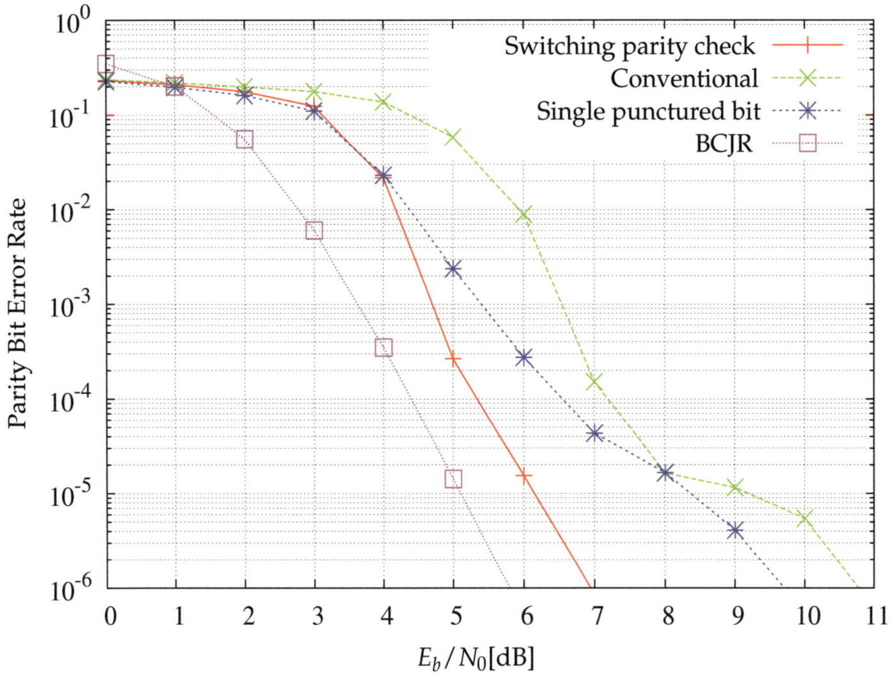


Fig. 16. Parity Bit Error Rate Performance of switching parity check method

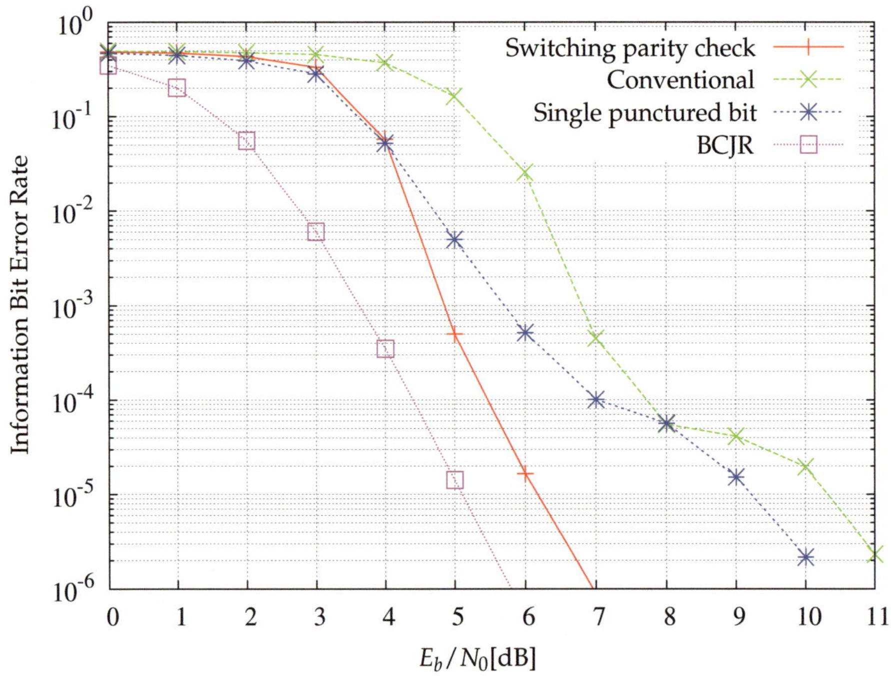


Fig. 17. Information Bit Error Rate Performance of switching parity check method

check method, it is necessary to notice that higher degree parity check polynomials No.2 and No.5 are used at only first iteration and after first iteration, higher degree parity check polynomials with degree $\nu = 15$ is used.

From those tables, for code rate 2/3, it can be seen that the number of operations of the single punctured bit method is 0.1 times of that of BCJR. For code rate 3/4, the number of operations of the switching parity check method is 0.2 times of that of BCJR.

For the both code rates, it can be seen that the number of operations of the proposed method is much less than that of BCJR.

| Code rate | Classification | $N_{add}$ | $N_{mult}$ | $N_{sp}$ | $N_{itr}$ | $N_{total}$ |
|---|---|---|---|---|---|---|
| 2/3 | Parity | 10 | 23 | 24 | 3.04 | 175 |
| | Info | 1 | 0 | 0 | 1 | |
| 3/4 | No.2 | 14 | 31 | 32 | 1 | 336 |
| | No.5 | 14 | 31 | 32 | 1 | |
| | $\nu = 15$ | 8 | 19 | 20 | 3.85 | |
| | Info | 1 | 0 | 0 | 1 | |

Table 6. Complexity of sum-product algorithm

| Code rate | $N_{add}$ | $N_{mult}$ | $N_{sp}$ | $N_{total}$ |
|---|---|---|---|---|
| 2/3, 3/4 | 640 | 1044 | 7 | 1691 |

Table 7. Complexity of BCJR

## 8. Conclusion

This paper proposes sum-product decoding methods for the punctured convolutional codes of wireless LAN. The wireless LAN standard include the punctured convolutional codes with code rate 2/3 and 3/4. This paper proposes to decode with the higher degree parity check polynomial that satisfies single punctured bit condition as the single punctured bit method. Single punctured bit condition is the condition to include single punctured bit in parity check equation. For code rate 2/3, the performance of the single punctured bit method is 1.28[dB] superior to that of the conventional method and only 0.98[dB] inferior to that of BCJR at bit error rate $10^{-5}$. For code rate 3/4, the single punctured bit method can not provide good performance. To improve the performance, this paper proposes following method as the switching parity check method. Only at first iteration, the higher degree parity check polynomial satisfying single punctured bit condition is used to decode and after first iteration, another higher degree parity check polynomial with lower $n_{fc}$ without single punctured bit condition is used to decode. For code rate 3/4, the performance of the switching parity check method is 4.16[dB] superior to that of the conventional method, 3.05[dB] superior to that of the single punctured bit method and only 1.06[dB] inferior to that of BCJR. Complexity of the single punctured bit method is 0.1 times of that of BCJR for code rate 2/3. For code rate 3/4, complexity of the switching parity check method is 0.2 times of that of BCJR. For the both code rates, complexity of the proposed method is much less than that of BCJR.

## 9. Acknowledgment

This work was supported by Japan Society for the Promotion of Science (JSPS) Grant-in-Aid for Scientific Research (C) 23560444.

## 10. References


Benedetto, S.; Divsalar, D.; Montorsi, G. & Pollara, F. (1998). Serial concatenation of interleaved codes: performance analysis, design, and iterative decoding. *IEEE Trans. Inf. Theory*, Vol. 44, No.3, 909-926

Berrou, C.; Glavieux, A. & Thitimajshima, P. (1993). Near shannon limit error-correcting coding and decoding : Turbo-codes (1). *IEEE International Conference on Communications (ICC93)*, 1064-1070

Berrou, C. & Glavieux, A. (1996). Near optimum error correcting coding and decoding: Turbo-codes. *IEEE Trans. Commun.*, Vol. 44, No. 10, 1261-1271

Di, C.; Proietti, D.; Telatar, E.; Richardson, T.; & Urbanke, R. (2002). Finite length analysis of low-density parity-check codes, *IEEE Trans. Inf. Theory*, Vol.48, No.6, 1570-1579

Douillard, Catherine; Jézéquel, Michel; Berrou, Claude; Picart, Annie; Didier, Pierre & Glavieux, Alain (1995). Iterative correction of intersymbol interference: Turbo-Equalization. *European Trans. Telecommun.*, Vol.6, No.5, 507–511

Gallager, R. G. (1963). *Low Density Parity Check Codes*, Cambridge, MA: MIT Press

Hagenauer, J.; Offer, E. & Papke, L. (1996). Iterative decoding of binary block and convolutional codes. *IEEE Trans. Inf. Theory*, Vol. 42, No. 2, 429-445

IEEE Computer Society (2007). Part11:Wireless LAN Medium Access Control (MAC) and Physical Layer (PHY) Specifications, *IEEE Std 802.11-2007*

Kschischang, Frank R.; Frey, Brendan J. & Loeliger, Hans-Andrea. Factor Graphs and the Sum-Product Algorithm, *IEEE Trans. Inf. Theory*, 498-519

Laot, Christophe; Glavieux, Alain & Labat, Joël (2001). Turbo equalization: adaptive equalization and channel decoding jointly optimized, *IEEE J. Selected Areas in Commun.*, Vol.19, No.9, 1744-1752

MacKay, D. J. C. (1999). Good error-correcting codes based on very sparse matrices, *IEEE Trans. Inf. Theory*, Vol. 45, No. 3, 399-431

Richardson, T. J. & Urbanke, R.L. (2001). The capacity of low-density parity-check codes under message-passing decoding, *IEEE Trans. Inf. Theory*, Vol. 47, No. 2, 599-618

Shohon, T.; Ogawa, Y. & Ogiwara, H. (2009a). Sum-Product decoding of convolutional codes, *The Fourth International Workshop on Signal Design and its Application in Communications (IWSDA'09)*, 64-67

Shohon, T.; Razi, F.; Ogawa, Y. & Ogiwara, H. (2009b). Sum-Product decoding of convolutional code for wireless LAN standard, *The Fourth International Workshop on Signal Design and its Application in Communications (IWSDA'09)*, 68-71

Shohon, T.; Razi, F. & Ogiwara, H. (2010). Sum-Product decoding of non-systematic convolutional codes, *Far East Journal of Electronics and Communications*, Vol.5, No.1, 25-35

2

# A MAC Throughput in the Wireless LAN

Ha Cheol Lee
*Dept. of Information and Telecom. Eng., Yuhan University, Bucheon City, Korea*

## 1. Introduction

Over the past few years, mobile networks have emerged as a promising approach for future mobile IP applications. With limited frequency resources, designing an effective MAC (Medium Access Control) protocol is a hot challenge. IEEE 802.11b/g/a/n networks are currently the most popular wireless LAN products on the market [1]. The conventional IEEE 802.11b and 802.11g/a specification provide up to 11 and 54 Mbps data rates, respectively. However, the MAC protocol that they are based upon is the same and employs a CSMA/CA (Carrier Sense Multiple Access/Collision Avoidance) protocol with binary exponential back-off. IEEE 802.11 DCF (Distributed Coordination Function) is the de facto MAC protocol for wireless LAN because of its simplicity and robustness [2,3]. Therefore, considerable research efforts have been put on the investigation of the DCF performance over wireless LAN [2]. With the successful deployment of IEEE 802.11a/b/g wireless LAN and the increasing demand for real-time applications over wireless, the IEEE 802.11n Working Group standardized a new MAC and PHY (Physical) layer specification to increase the bit rate to be up to 600 Mbps [3]. The throughput performance at the MAC layer can be improved by aggregating several frames before transmission [3]. Frame aggregation not only reduces the transmission time for preamble and frame headers, but also reduces the waiting time during CSMA/CA random backoff period for successive frame transmissions. The frame aggregation can be performed within different sub-layers. In 802.11n, frame aggregation can be performed either by A-MPDU (MAC Protocol Data Unit Aggregation) or A-MSDU (MAC Service Data Unit Aggregation). Although frame aggregation can increase the throughput at the MAC layer under ideal channel conditions, a larger aggregated frame will cause each station to wait longer before its next chance for channel access. Under error-prone channels, corrupting a large aggregated frame may waste a long period of channel time and lead to a lower MAC efficiency [4]. On the other hand, wireless LAN mobile stations that are defined as the stations that access the LAN while in motion are considered in this chapter. The previous paper analyzed the IEEE 802.11b/g/n MAC performance for wireless LAN with fixed stations, not for wireless LAN with mobile stations [5, 6, 7, 8, 9, 10]. On the contrary, Xi Yong [11] and Ha Cheol Lee [12] analyzed the MAC performance for IEEE 802.11 wireless LAN with mobile stations, but considered only IEEE 802.11 and 802.11g/a wireless LAN specification. So, this chapter summarizes all the reference papers and analyzes the IEEE 802.11b/g/a/n MAC performance for wireless LAN with fixed and mobile stations. In other words, we will present the analytical evaluation of saturation throughput with bit errors appearing in the transmitting channel. In Section 2, wireless LAN history and standards are reviewed. In section 3, wireless LAN access network is reviewed. IEEE 802.11b/g/a/n/ac/ad PHY and MAC layer are reviewed in Section 4. In Section 5, frame error rate of wireless channel and the DCF saturation throughput are theoretically derived. Finally, it is concluded with Section 6.

## 2. Wireless LAN history and standards

Standards in the IEEE project 802 target the PHY layer and MAC layer. When wireless LAN was first conceived, it seemed that it would be just another PHY of one of the available standards. The first candidate considered for this was IEEE's most prominent standard 802.3 (Ethernet). However, it soon became obvious that the radio medium is very different from the well-behaved wire. Due to tremendous attenuation even over short distances, collisions cannot be detected. Hence, 802.3's CSMA/CD (Carrier Sense Multiple Access/Collision Detection) could not be applied. The next candidate standard to be considered was 802.4. Its coordinated medium access, the token bus concept, was believed to be superior to 802.3's contention-based scheme. Hence, WLAN began as 802.4L. However, already in 1990 it was obvious that token handling in radio networks was difficult. The standardization body realized that a wireless communication standard would need its own very unique MAC. Finally, on March 21, 1991, project 802.11 was approved. The first 802.11 standard was published in 1997. At the PHY layer it provides three solutions: a FHSS (Frequency Hopping Spread Spectrum) and a DSSS (Direct Sequence Spread Spectrum) PHY in the unlicensed 2.4 GHz band, and an infrared PHY at 316–353 THz. Although all three provide a basic data rate of 1 Mb/s with an optional 2 Mb/s mode, commercial infrared implementations do not exist. Similar to 802.3, basic 802.11 MAC operates according to a listen-before-talk scheme, and is known as the DCF. It implements CSMA/CA rather than collision detection as in 802.3. Indeed, as collision cannot be detected in the radio environment, 802.11 waits for a backoff interval before each frame transmission rather than after collisions. In addition to DCF, the original 802.11 standard specifies an optional scheme that depends on a central coordination entity, the PCF (Point Coordination Function). This function uses the so called PC (Point Coordinator) that operates during the so-called contention-free period. The latter is a periodic interval during which only the PC initiates frame exchanges via polling. However, the PCF's poor robustness against hidden nodes resulted in negligible adoption by manufacturers. Having published its first 802.11 standard in 1997, the WG (Working Group) received feedback that many products did not provide the degree of compatibility customers expected. As an example, often the default encryption scheme, called WEP (Wired Equivalent Privacy), would not work between devices of different vendors. This need for a certification program led to the foundation of the WECA (Wireless Ethernet Compatibility Alliance) in 1999, renamed the WFA (Wi-Fi Alliance) in 2003. Wi-Fi certification has become a well-known certification program that has significant market impact. The tremendous success in the market and the perceived shortcomings of the base 802.11 standard provided a basis and impetus for a prolific program of improvements and extensions. This has led to revisions of the draft, driven by a complete alphabet of amendments. It is the purpose of this article to review this process and explain both the contents of these amendments and their interrelation. In the following we first describe the changes made to the PHY layer and then turn to the improvements to the MAC layer. In both, we make a distinction between what has already been accepted and what is currently in the process of being standardized [2].

| Standard | Spectrum | Maximum physical rate | Layer 2 data rate | Tx | Compatible with | Major disadvantage | Major advantage |
|---|---|---|---|---|---|---|---|
| 802.11n | 2.4/5 GHz | 600 Mbps | 100 Mbps | MIMO OFDM | 802.11b/g/a | Difficult to implement | Highest bit rate |
| 802.11b | 2.4 GHz | 11 Mbps | 6-7 Mbps | DSSS | 802.11 | Bit rate too low for many emerging applications | Widely deployed, higher range |
| 802.11g | 2.4 GHz | 54 Mbps | 32 Mbps | OFDM | 802.11/ 802.11b | Limited number of collocated WLANs | Higher bit rate in 2.4 GHz spectrum |
| 802.11a | 5.0 GHz | 54 Mbps | 32 Mbps | OFDM | None | Smallest range of all 802.11 standards | Higher bit rate in less-crowded spectrum |

Table 1. Wireless LAN products on the market [1]

### 2.1 PHY related amendments

Although not interoperable, the DSSS and FHSS PHY initially seemed to have equal chances in the market. The FHSS PHY even had a duplicate in the HomeRF group that aimed at integrated voice and data services. This used plain 802.11 with FHSS for data transfer, complemented with a protocol for voice that was very similar to the Digital Enhanced Cordless Telecommunications standard. Neither HomeRF nor 802.11 saw FHSS extensions, although plans for a second-generation HomeRF existed that targeted at 10 Mb/s. In contrast, the high-rate project 802.11b was started in December 1997 and boosted the data rates of the DSSS PHY to 11 Mb/s. This caused 802.11b to ultimately supersede FHSS, including HomeRF, in the market. Figure 1 provides an overview of the 802.11 PHY amendments and their dependencies [2].

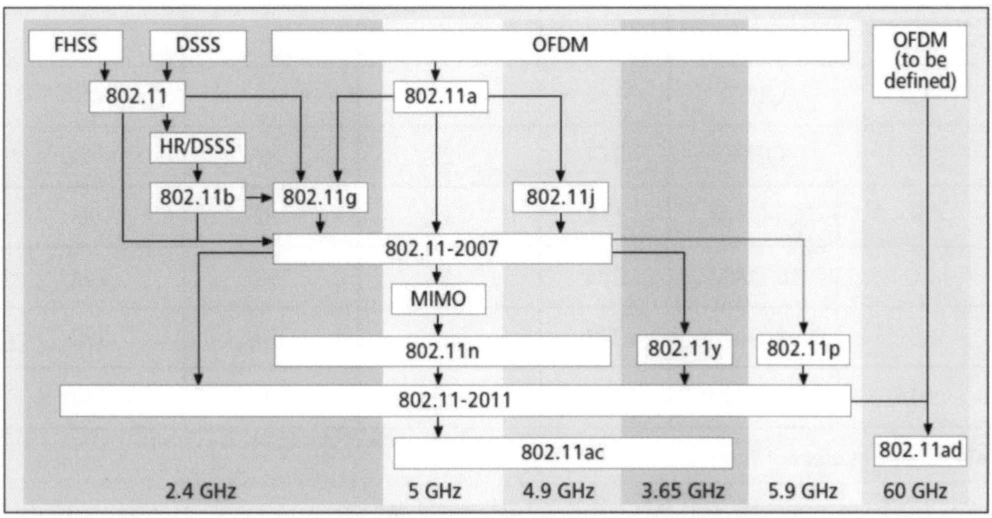


Fig. 1. The 802.11 PHY layer amendments and their dependencies [2]

### 2.1.1 802.11a/g

The first extension project, 802.11a, started in September 1997. It added an OFDM (Orthogonal Frequency Division Multiplexing) PHY that supports up to 54 Mb/s data rate. Since 802.11a operates in the 5 GHz band, communication with plain 802.11 devices is impossible. This lack of interoperability led to the formation of 802.11g, which introduced the benefits of OFDM to the 2.4 GHz band. As 802.11g's extended rate PHY provides DSSS-compatible signaling, an easy migration from 802.11 to 802.11g devices became possible. During the standardization process, a single manufacturer already sold pre-802.11g chipsets. With its proprietary PBCC (Packet Binary Convolutional Code), additional data rates of 22 Mb/s and 33 Mb/s were supported. Today rarely applied, PBCC set a de facto standard and became an optional MCS (Modulation and Coding Scheme) of 802.11g. To comply with the European regulatory requirements for the 5 GHz band, 802.11h was introduced at the end of 2003. While in the United States the FCC describes absolute radio output power limits, in Europe antenna gain must not be used for transmission. Furthermore, satellite uplink and radar stations must be secured from interference. Therefore, 802.11h defines MAC mechanisms for DFS (Dynamic Frequency Selection) and TPC (Transmit Power Control), which we explain in the MAC section. Ratified in 2004, 802.11j describes the necessary means to comply with Japanese regulatory requirements for the operation of 802.11 equipment in the 4.9 GHz and 5 GHz frequency bands. Besides requirements on medium access discussed in the next section, 802.11j is the first amendment that defines PHY operation with 10 MHz bandwidth in addition to the formerly preferred 20 MHz channelization.

| Data Rate (Mbits/sec) | Modulation | Coding Rate (R) | Coding Bits Per Subcarrier ($N_{BPSC}$) | Coded Bits per OFDM symbol ($N_{CBPS}$) | Data Bits Per OFDM Symbol ($N_{DBPS}$) |
|---|---|---|---|---|---|
| 6 | BPSK | 1/2 | 1 | 48 | 24 |
| 9 | BPSK | 3/4 | 1 | 48 | 36 |
| 12 | QPSK | 1/2 | 2 | 96 | 48 |
| 18 | QPSK | 3/4 | 2 | 96 | 72 |
| 24 | 16-QAM | 1/2 | 4 | 192 | 96 |
| 36 | 16-QAM | 3/4 | 4 | 192 | 144 |
| 48 | 64-QAM | 2/3 | 6 | 288 | 192 |
| 54 | 64-QAM | 3/4 | 6 | 288 | 216 |

Table 2. Parameters of the IEEE 802.11a physical layer

While IEEE 802.11b uses only DSSS technology, IEEE 802.11g uses DSSS, OFDM, or both at the 2.4 GHz ISM band to provide high data rates of up to 54 Mb/s. Combined use of both

DSSS and OFDM is achieved through the provision of four different physical layers. These layers, defined in the standard as ERPs (Extended Rate Physicals), coexist during a frame exchange, so the sender and receiver have the option to select and use one of the four layers as long as they both support it. The four different physical layers defined in the IEEE 82.11g standard are the following :

- *ERP-DSSS/CCK*: This is the old physical layer used by IEEE 802.11b. DSSS technology is used with CCK modulation. The data rates provided are those of IEEE 802.11b.
- *ERP-OFDM*: This is a new physical layer, introduced by IEEE 802.11g. OFDM is used to provide IEEE 802.11a data rates at the 2.4 GHz band.
- *ERP-DSSS/PBCC*: This physical layer was introduced in IEEE 802.11b and provides the same data rates as the DSSS/CCK physical layer. It uses DSSS technology with the PBCC coding algorithm. IEEE 802.11g extended the set of data rates by adding those of 22 and 33 Mb/s.
- *DSSS-OFDM*: This is a new physical layer that uses a hybrid combination of DSSS and OFDM. The packet physical header is transmitted using DSSS, while the packet payload is transmitted using OFDM. The scope of this hybrid approach is to cover interoperability aspects, as explained later. From the above four physical layers, the first two are mandatory; every IEEE 802.11g device must support them. The other two are optional. Column 2 of Table 3 summarizes the supported data rates for the different physical layers of the IEEE 802.11g specification.

| Physical layer | Supported rates (Mb/s) | PLCP preamble + header delay | | PLCP preamble + header length | |
|---|---|---|---|---|---|
| | | Long | Short | Long | Short |
| ERP-DSSS (mandatory) | 1, 2, 5.5, 11 | 192 μs | 96 μs | 192 bits | 120 bits |
| ERP-OFDM (mandatory) | 6, 9, 12, 18, 24, 36, 48, 54 | 20 μs | | 40 bits | |
| ERP-PBCC (optional) | 1, 2, 5.5, 11, 22, 33 | 192 μs | 96 μs | 192 bits | 120 bits |
| DSSS-OFDM (optional) | 6, 9, 12, 18, 24, 36, 48, 54 | 192 μs | 96 μs | 192 bits | 120 bits |

Table 3. Parameters of the different IEEE 802.11g physical layers [8]

### 2.1.2 802.11n

As the first project whose targeted data rate is measured on top of the MAC layer, 802.11n provides user experiences comparable to the well known Fast Ethernet (802.3u). Far beyond the minimum requirements that were derived from its wired paragon's maximum data rate of 100 Mb/s, 802.11n delivers up to 600 Mb/s. Its most prominent feature is MIMO capability. A flexible MIMO (Multiple Input Multiple Output) concept allows for arrays of up to four antennas that enable spatial multiplexing or beam forming. Its most debated innovation is the usage of optional 40 MHz channels. Although this feature was already being used as a proprietary extension to 802.11a and 802.11g chipsets, it caused an extensive discussion on neighbor friendly behavior. Especially for the 2.4 GHz band, concerns were raised that 40 MHz operation would severely affect the performance of existing 802.11, Bluetooth (802.15.1), ZigBee (802.15.4), and other devices. The development of a compromise, which disallows 40 MHz channelization for devices that cannot detect 20 MHz-only devices, prevented ratification of 802.11n until September 2009. As a consequence of 20/40 MHz operation and various antenna configurations, 802.11n defines a total of 76 different MCSs. Since several of them provide similar data rates, WFA's certification program decides the MCSs finally used in the market. 802.11n's PHY enhancements are supported by medium access enhancements we introduce in the MAC section.

| MCS Index | Modulation | Coding Rate | Spatial Streams | 802.11n Data Rate (Mbps) | | | |
|---|---|---|---|---|---|---|---|
| | | | | 20-MHz | | 40-MHz | |
| | | | | L-GI | S-GI | L-GI | S-GI |
| 0 | BPSK | 1/2 | 1 | 6.5 | 7.2 | 13.5 | 15 |
| 1 | QPSK | 1/2 | 1 | 13 | 14.4 | 27 | 30 |
| 2 | QPSK | 3/4 | 1 | 19.5 | 21.7 | 40.5 | 45 |
| 3 | 16-QAM | 1/2 | 1 | 26 | 28.9 | 54 | 60 |
| 4 | 16-QAM | 3/4 | 1 | 39 | 43.3 | 81 | 90 |
| 5 | 64-QAM | 2/3 | 1 | 52 | 57.8 | 108 | 120 |
| 6 | 64-QAM | 3/4 | 1 | 58.5 | 65 | 122 | 135 |
| 7 | 64-QAM | 5/6 | 1 | 65 | 72.2 | 135 | 150 |
| 8 | BPSK | 1/2 | 2 | 13 | 14.4 | 27 | 30 |
| 9 | QPSK | 1/2 | 2 | 26 | 28.9 | 54 | 60 |
| 10 | QPSK | 3/4 | 2 | 39 | 43.3 | 81 | 90 |
| 11 | 16-QAM | 1/2 | 2 | 52 | 57.8 | 108 | 120 |
| 12 | 16-QAM | 3/4 | 2 | 78 | 86.7 | 162 | 180 |
| 13 | 64-QAM | 2/3 | 2 | 104 | 116 | 216 | 240 |
| 14 | 64-QAM | 3/4 | 2 | 117 | 130 | 243 | 270 |
| 15 | 64-QAM | 5/6 | 2 | 130 | 144 | 270 | 300 |
| 16 | BPSK | 1/2 | 3 | 19.5 | 21.7 | 40.5 | 45 |
| 17 | QPSK | 1/2 | 3 | 39 | 43.3 | 81 | 90 |
| 18 | QPSK | 3/4 | 3 | 58.5 | 65 | 121.5 | 135 |
| 19 | 16-QAM | 1/2 | 3 | 78 | 86.7 | 162 | 180 |
| 20 | 16-QAM | 3/4 | 3 | 117 | 130 | 243 | 270 |
| 21 | 64-QAM | 2/3 | 3 | 156 | 173.3 | 324 | 360 |
| 22 | 64-QAM | 3/4 | 3 | 175.5 | 195 | 364.5 | 405 |
| 23 | 64-QAM | 5/6 | 3 | 195 | 216.7 | 405 | 450 |
| 24 | BPSK | 1/2 | 4 | 26 | 28.9 | 54 | 60 |
| 25 | QPSK | 1/2 | 4 | 52 | 57.8 | 108 | 120 |
| 26 | QPSK | 1/2 | 4 | 78 | 86.7 | 162 | 180 |
| 27 | 16-QAM | 1/2 | 4 | 104 | 115.6 | 216 | 240 |
| 28 | 16-QAM | 3/4 | 4 | 156 | 173.3 | 324 | 360 |
| 29 | 64-QAM | 2/3 | 4 | 208 | 231.1 | 432 | 480 |
| 30 | 64-QAM | 3/4 | 4 | 234 | 260 | 486 | 540 |
| 31 | 64-QAM | 5/6 | 4 | 260 | 288.9 | 540 | 600 |

Table 4. Parameters of the IEEE 802.11n physical layer, MCS Rates 0-31 [13]

### 2.1.3 802.11ac/ad

802.11ac and 802.11ad develop amendments that fulfill the ITU's (International Telecommunication Union's) requirements on proposals for the IMT Advanced standard. Both target greater than 1 Gb/s throughput, but while 802.11ac considers the traditional Wireless LAN frequencies below 6 GHz, 802.11ad competes with the Wireless Personal Area Network TG (Task Group) 802.15.3c, standard ECMA 387, and the Wireless Gigabit Alliance on the 60 GHz frequency spectrum. Due to their premature stage, both TGs are still in the process of collecting input and specific proposals from their members. At the moment of writing this article, 802.11ad has already started defining some additional requirements regarding range (at least 10 m at 1 Gb/s), seamless session transfer of an active session from the 60 GHz band to the 2.4/5 GHz band and vice versa, coexistence with other systems in the band such as 802.15.3c, and support for uncompressed video requirements such as data rate, packet loss ratio, and delay.

## 2.2 MAC related amendments

A key element to the 802.11 success is its simple MAC operation based on the DCF protocol. This scheme has proven to be robust and adaptive to varying conditions, able to cover most needs sufficiently well. Following the trends visible from the wired Ethernet, 802.11's success is mainly based on overprovisioning of its capacity. The available data rate was sufficient to cover the original best effort applications, so complex resource scheduling and management algorithms were unnecessary.However, this may change in the future. Because of the growing popularity of 802.11, WLANs are expected to reach their capacity limits. Moreover, applications like voice and video streaming pose different demands for quality of service. Therefore, traffic differentiation and network management might become inevitable. In the following we explain 802.11 MAC related extensions of the amendments introduced in the previous section and those shown in Fig. 2 [2].

### 2.2.1 802.11e

The original project goal of 802.11e, approved at the end of March 2000, foresaw general enhancements of the WLAN standard. Efficiency improvements, support for quality of service (QoS), and security enhancements were its key elements. However, already in 2001, the 802.11 frame encryption algorithm WEP was broken by an attack. Thus, security enhancements were displaced to a new TG called 802.11i. After intensive discussions, 802.11e was finally approved in 2005 to support QoS. As a new medium access scheme, 802.11e provides the HCF (Hybrid Coordination Function), where *hybrid* relates to HCF's two MAC protocol versions with centralized and distributed control, respectively. The first is implemented by HCF HCCA (HCF Controlled Channel Access), an improved variant of the PCF requiring a central coordination instance that schedules medium access. Until today no device implementing HCCA is known to exist in the market. EDCA is HCF's second MAC protocol. While DCF does not differentiate between traffic with different QoS needs, EDCA (Enhanced Distributed Channel Access) provides support for four traffic categories: voice, video, best effort, and background with different rules to access the wireless medium. Accordingly, EDCA enables service differentiation. Both centralized and distributed MAC protocols change the medium sharing rules. Without 802.11e, a WLAN provides per packet fairness: regardless of the actual frame transmission duration, devices back off after every single frame. In contrast, duration of all

HCCA and EDCA frame exchanges is bound by the TXOP (Transmission Opportunity) limit. Thus, devices share time slices of the wireless medium. Those that use faster MCSs may exchange multiple frames after a single successful contention and consequently achieve higher throughput. Derived from EDCA, WFA has successfully branded and introduced to the market an EDCA variant called WMM (Wi-Fi MultiMedia). WMM incorporates a subset of functions from 802.11e draft 6 (November 2003). As the final 802.11e and WMM specifications differ, some members of the 802.11 initiated a QoS Enhancement SG (Study Group) in May 2007. Its intended goal was an adaptation of the 802.11e amendment to the WMM specification. However, a project could never be approved, and the SG was dissolved in November 2007.

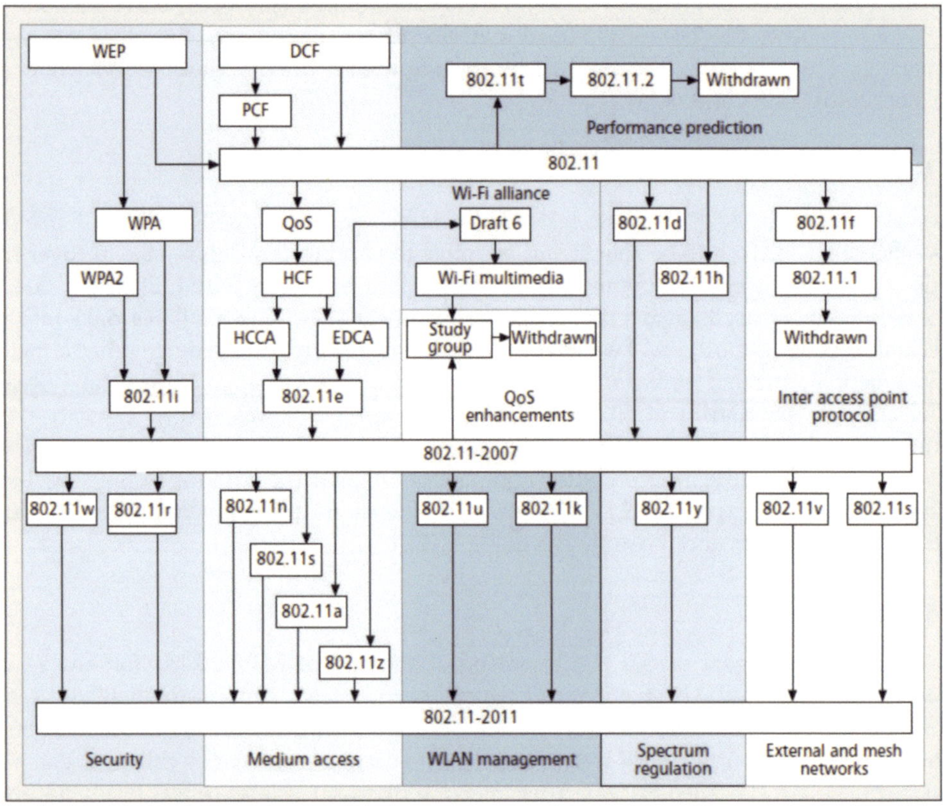


Fig. 2. The 802.11 MAC layer amendments [2]

## 3. Wireless LAN access network

This section shows infrastructure-based and ad hoc-based operation of wireless access architecture in the 802.11b/a/g/n-based mobile LAN. The protocols of the various layers are called the protocol stack. The TCP/IP protocol stack consists of five layers: the physical, data link, network, transport and application layers. This section is focused on physical layer and data link layer which consists of MAC and LLC (Logical Link Control) sub-layers. An ad hoc network might be formed when people with laptops get together and want to exchange data in the absence of a centralized AP (Access Point). Wireless LAN topology is ad hoc-based or infrastructure-based as shown in Fig. 3. The ad hoc-based topology shows that each user in the wireless network communicates directly with all others without a backbone network. Infrastructure-based topology shows that all wireless users transmit to an AP to communicate with users on the wired or wireless LAN. IEEE 802.11 operates in the 2.4 GHz band and supports data rates 1 Mbps to 2 Mbps. IEEE 802.11b uses DSSS (Direct Sequence Spread Spectrum) but supports data rates of up to 11 Mbps. The modulation scheme employed is called CCK (Complementary Code Keying). The operating frequency range is 2.4 GHz and hence can interfere with some home appliances. IEEE 802.11g achieves very high data rates compared to IEEE 802.11b and uses the 2.4 GHz frequency band. An IEEE 802.11b client can operate with an 802.11g AP. IEEE 802.11a equipment is more expensive and consumes more power, as it uses OFDM (Orthogonal Frequency Division Multiplexing). OFDM uses 12 orthogonal channels in the 5 GHz range. The frequency channels are nonoverlapping. The achievable data rates are 6, 9, 12, 18, 24, 36, 48 and 54 Mbps. IEEE 802.11a and 802.11b can operate next to each other without any interference. Fig. 4 shows the IEEE 802.11b/a/g/n-based physical and MAC layer protocol stack.

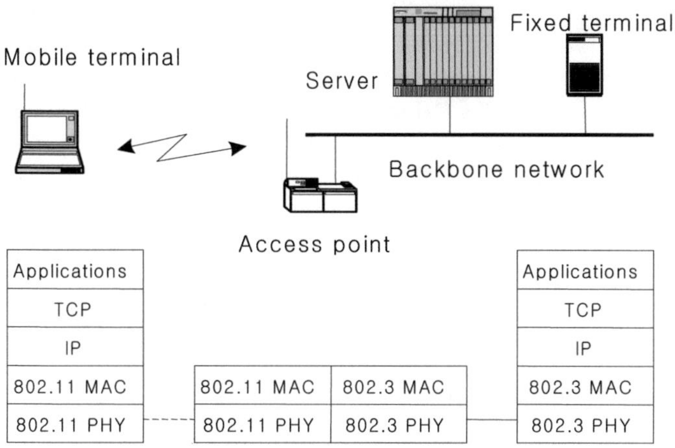


(a) Infrastructure-based wireless LAN

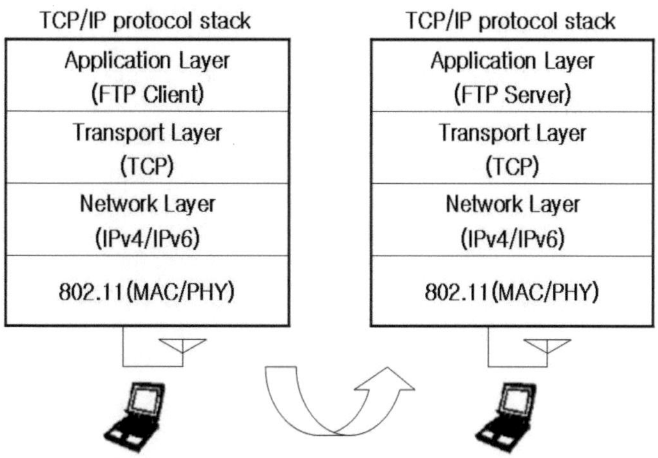


(b) Ad-hoc mode operation in the wireless LAN

Fig. 3. Protocol stack in the IEEE 802.11 wireless LAN [12]

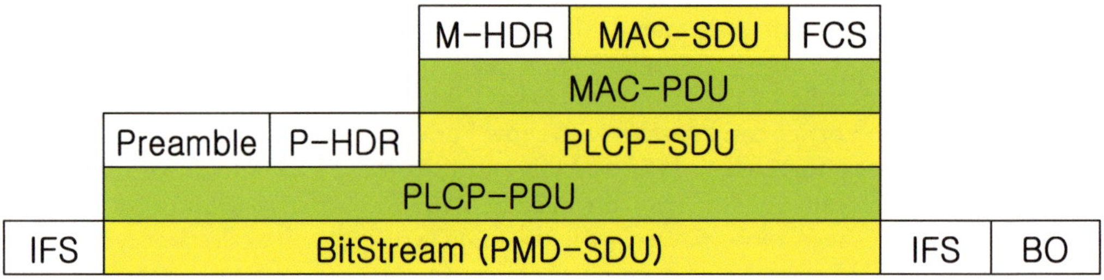


Fig. 4. Protocol stack of physical and MAC layer [12]

IEEE 802.11 protocol stack consists of MAC layer and PHY layer. When a network layer pushes a user packet down to the MAC layer as a MAC-SDU (MAC-Service Data Unit), overheads are added to the MAC layer and MAC-PDU (MAC-Protocol Data Unit) is created. The PHY layer is divided into a PLCP (Physical Layer Convergence Protocol) sublayer and a PMD (Physical Medium Dependent) sublayer. In this PHY layer, the same procedure as MAC layer is also executed. IEEE 802.11 MAC layer uses an 802.11 PHY layer, such as 802.11a/b/g, to perform the tasks such as carrier sensing, transmission, and reception of 802.11 frames. With regards to the MAC layer, the functional specifications are essentially the same for all of them with minor differences.

## 4. Wireless LAN PHY/MAC layer

### 4.1 IEEE 802.11b/a/g PHY/MAC layer

When a higher layer pushes a user packet down to the MAC layer as a MAC-SDU, the MAC layer header (M-HDR) and trailer (FCS) are added before and after the MSDU, respectively and form a MAC-PDU. The PHY layer is again divided into a PLCP sub-layer and a PMD sub-layer. Similarly the PLCP preamble and PLCP header (P-HDR) are attached to the MPDU at the PLCP sub-layer. Different IFS (Inter Frame Space)s are added depending on the type of MPDU. IEEE 802.11a operates in the 5 GHz band and uses OFDM. The achievable data rates are 6, 9, 12, 18, 24, 36, 48, and 54 Mbps. 802.11g uses DSSS, OFDM, or both at the 2.4 GHz ISM band to provide high data rates of up to 54 Mbps. 802.11g device can operate with an 802.11b device. Combined use of both DSSS and OFDM is achieved through the provision of four different physical layers. The four different physical layers defined in the 802.11g standards are ERP-DSSS/CCK, ERP-OFDM, ERP-DSSS/PBCC and DSSS-OFDM. The standards that support the highest data rate of 54 Mbps are ERP-OFDM and DSSS-OFDM. ERP-OFDM is a new physical layer in IEEE 802.11g and OFDM is used to provide IEEE 802.11a data rates at the 2.4 GHz band. DSSS-OFDM is a new physical layer that uses a hybrid combination of DSSS and OFDM. The packet physical header is transmitted using DSSS, while the packet payload is transmitted using OFDM. Basic access scheme is CSMA/CA mechanism. The SIFS (Short Inter-Frame Space) and the slot time are determined by the physical layer. DIFS (Distributed Inter-Frame Space) is defined based on the above two intervals.

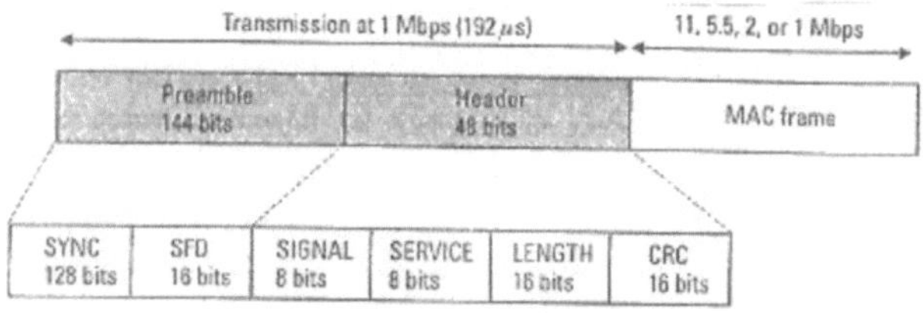


(a) 802.11b ERP-OFDM frame

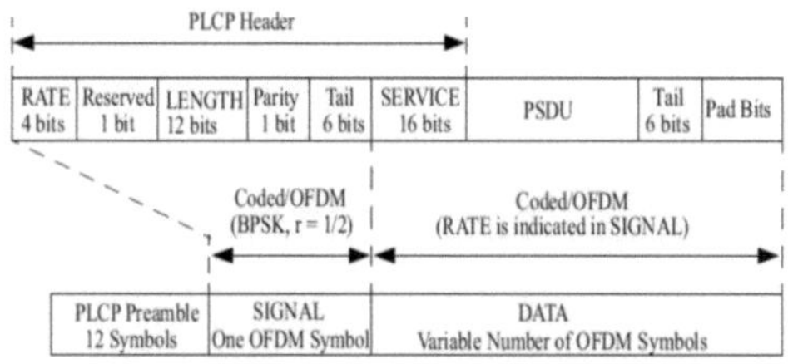


(b) 802.11a and 802.11g ERP-OFDM frame

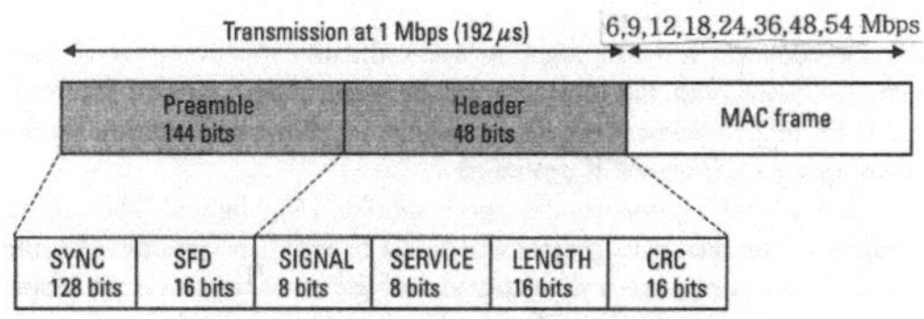


(c) 802.11g DSSS-OFDM frame

Fig. 5. Frame structure of IEEE 802.11b/a/g-based wireless LAN [12]

IEEE 802.11 MAC protocol supports the DCF and the PCF . The DCF uses the CSMA/CA mechanism for contention-based access, while the PCF provides contention-free access. The two modes are used alternately in time. IEEE 802.11 MAC protocol defines five timing intervals. Two of them are the SIFS and the slot time that are determined by the physical layer. The other three intervals are the PIFS (Priority InterFrame Space), DIFS and EIFS (Extended InterFrame Space) that are defined based on the above two intervals. But the PCF is restricted to infrastructure network configurations. Therefore, the DCF is widely assumed under the consideration of ad hoc-based wireless LAN. Fig. 6 shows two access schemes. IEEE 802.11 DCF stations access the channel via a basic access method or the four-way handshaking access method with an additional RTS/CTS message exchange. In the basic access method, the CSMA mechanism is applied. Stations wait for the channel to be idle for a DIFS period of time and then execute backoff for data transmission. Stations choose a random number between 0 and CW (Contention Window)-1 with equal probability as a backoff timer. When the backoff timer reaches zero, the data frame is transmitted. The receiver replies an ACK message upon successfully receiving a data packet. In the four-way handshaking access method, when the backoff timer of station reaches zero, the station first transmits a RTS frame. Upon receiving the RTS frame, the receiver replies with a CTS frame after a SIFS period. Once the RTS/CTS is exchanged successfully, the sender then transmits its data frame. The RTS and CTS frames carry a duration field, information of time interval to transmit the packet. Any station receiving RTS or CTS frames can read the duration field information. That information is then used to update a NAV (Network Allocation Vector) value that indicates to each station the amount of time that remains before the channel will become idle. Therefore, a station detecting the RTS and CTS frames suitably delays further transmission, and thus avoids collision. The NAV is thus referred to as a virtual carrier sensing mechanism. The main purpose of the RTS/CTS handshaking is to resolve the so-called hidden node problem.

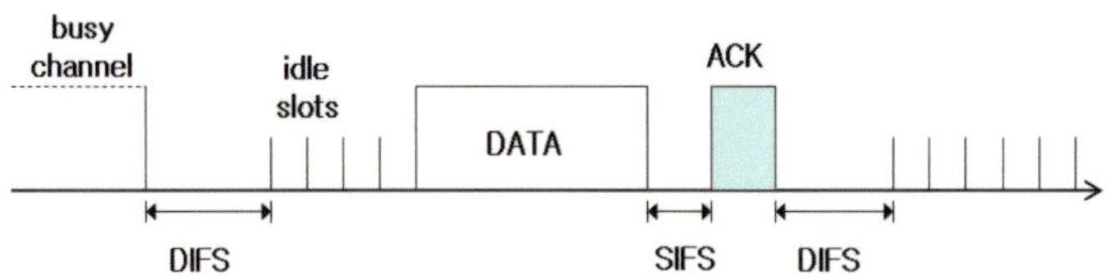


(a) Basic access method

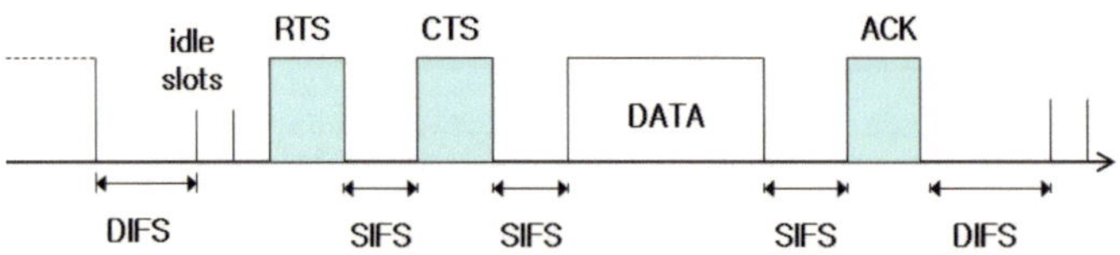


(b) Four-way handshaking access method

Fig. 6. IEEE 802.11 DCF channel access mechanism [14]

| Parameter | 802.11b | 802.11g | 802.11a |
|---|---|---|---|
| $T_{slot}$ | 20μs | 9μs/20μs | 9μs |
| T | 1μs | 1μs | 1μs |
| $T_P$ | 144μs | 16μs/144μs | 16μs |
| $CW_{min}$ | 31 | 31/15 | 15 |
| $T_{PHY}$ | 48μs | 4μs/48μs | 4μs |
| $T_{SYM}$ | N/A | N/A, 4μs | 4μs |
| $T_{DIFS}$ | 50μs | 50μs/34μs | 34μs |
| $T_{SIFS}$ | 10μs | 16μs | 16μs |

Table 5. Parameters of IEEE 802.11b/g/a [6, 7, 8]

| MAC scheme | Transmission | $T_{DIFS}$ | $T_{SIFS}$ | $T_{BO}$ | $T_{RTS}$ | $T_{CTS}$ | $T_{ACK}$ | $T_{DATA}$ (MSDU : bytes) |
|---|---|---|---|---|---|---|---|---|
| CSMA/CA | DSSS-1 | 50 | 10 | 310 | N/A | N/A | 304 | 192+8×(34+MSDU)/1 |
| | DSSS-2 | 50 | 10 | 310 | N/A | N/A | 304 | 192+8×(34+MSDU)/2 |
| | HR-5.5 | 50 | 10 | 310 | N/A | N/A | 304 | 192+8×(34+MSDU)/5.5 |
| | HR-11 | 50 | 10 | 310 | N/A | N/A | 304 | 192+8×(34+MSDU)/11 |
| | OFDM-6 | 34 | 9 | 67.5 | N/A | N/A | 44 | 20+4×[(16+6+8×(34+MSDU))/24] |
| | OFDM-12 | 34 | 9 | 67.5 | N/A | N/A | 32 | 20+4×[(16+6+8×(34+MSDU))/48] |
| | OFDM-24 | 34 | 9 | 67.5 | N/A | N/A | 28 | 20+4×[(16+6+8×(34+MSDU))/96] |
| | OFDM-54 | 34 | 9 | 67.5 | N/A | N/A | 24 | 20+4×[(16+6+8×(34+MSDU))/216] |
| RTS/CTS | DSSS-1 | 50 | 10×3 | 310 | 352 | 304 | 304 | 192+8×(34+MSDU)/1 |
| | DSSS-2 | 50 | 10×3 | 310 | 352 | 304 | 304 | 192+8×(34+MSDU)/2 |
| | HR-5.5 | 50 | 10×3 | 310 | 352 | 304 | 304 | 192+8×(34+MSDU)/5.5 |
| | HR-11 | 50 | 10×3 | 310 | 352 | 304 | 304 | 192+8×(34+MSDU)/11 |
| | OFDM-6 | 34 | 9×3 | 67.5 | 52 | 44 | 44 | 20+4×[(16+6+8×(34+MSDU))/24] |
| | OFDM-12 | 34 | 9×3 | 67.5 | 36 | 32 | 32 | 20+4×[(16+6+8×(34+MSDU))/48] |
| | OFDM-24 | 34 | 9×3 | 67.5 | 28 | 28 | 28 | 20+4×[(16+6+8×(34+MSDU))/96] |
| | OFDM-54 | 34 | 9×3 | 67.5 | 24 | 24 | 24 | 20+4×[(16+6+8×(34+MSDU))/216] |

Table 6. Delay compents for different MAC schemes (unit : μs) [7]

### 4.2 IEEE 802.11n PHY/MAC layer

The key requirement that drove most of the development in 802.11n is the capability of at least 100 Mb/s MAC throughput. Considering that the typical throughput of 802.11a/g is 25 Mb/s (with a 54 Mb/s PHY data rate), this requirement dictated at least a fourfold increase in throughput. Defining the requirement as MAC throughput rather than PHY data rate forced developers to consider the difficult problem of improving MAC efficiency. The inability to achieve a throughput of 100 Mb/s necessitated substantial improvements in MAC efficiency when designing the 802.11n MAC. Two basic concepts are employed in 802.11n to increase the PHY data rates: MIMO and 40 MHz bandwidth channels. Increasing from a single spatial stream and one transmit antenna to four spatial streams and four antennas increases the data rate by a factor of four. The term *spatial stream* is defined in the 802.11n standard as one of several bitstreams that are transmitted over multiple spatial dimensions created by the use of multiple antennas at both ends of a communications link. However, due to the inherent increased cost associated with increasing the number of antennas, modes that use three and four spatial streams are optional. And to allow for handheld devices, the two spatial streams mode is only mandatory in an AP. 40 MHz bandwidth channel operation is optional in the standard due to concerns regarding interoperability between 20 and 40 MHz bandwidth devices, the permissibility of the use of 40 MHz bandwidth channels in the various regulatory domains, and spectral efficiency. However, the 40 MHz bandwidth channel mode has become a core feature due to the low cost of doubling the data rate from doubling the bandwidth. Almost all 802.11n products on the market feature a 40 MHz mode of operation. Other minor modifications were also made to the 802.11a/g waveform to increase the data rate. The highest encoder rate in 802.11a/g is 3/4. This was increased to 5/6 in 802.11n for an 11 percent increase in data rate. With the improvement in RF (Radio Frequency) technology, it was demonstrated that two extra frequency subcarriers could be squeezed into the guard band on each side of the spectral waveform and still meet the transmit spectral mask. This increased the data rate by 8 percent over 802.11a/g. Lastly, the waveform in 802.11a/g and mandatory operation in 802.11n contains an 800 ns guard interval between each OFDM symbol. An optional mode was defined with a 400 ns guard interval between each OFDM symbol to increase the data rates by another 11 percent. Another functional requirement of 802.11n was interoperability between 802.11a/g and 802.11n. The TG decided to meet this requirement in the physical layer by defining a waveform that was backward compatible with 802.11a and OFDM modes of 802.11g. The preamble of the 802.11n mixed format waveform begins with the preamble of the 802.11a/g waveform. This includes the 802.11a/g short training field, long training field, and signal field. This allows 802.11a/g devices to detect the 802.11n mixed format packet and decode the signal field. Even though the 802.11a/g devices will not be able to decode the remainder of the 802.11n packet, they will be able to properly defer their own transmission based on the length specified in the signal field. The remainder of the 802.11n Mixed format waveform includes a second short training field, additional long training fields, and additional signal fields followed by the data. These new fields are required for MIMO training and signaling of the many new modes of operation. To ensure backward compatibility between 20 MHz bandwidth channel devices (including 802.11n and 802.11a/g) and 40 MHz bandwidth channel devices, the preamble of the 40 MHz waveform is identical to the 20 MHz waveform and is repeated on the two adjacent 20 MHz bandwidth channels that form the 40 MHz bandwidth channel. This allows 20 MHz bandwidth devices on either adjacent channel to decode the signal field and properly defer transmission. The preamble in 802.11a has a length of 20 μs; with the additional training and signal fields, the 802.11n mixed format packet has a preamble with a length of 36 μs for one spatial stream up to 48 μs for four spatial streams. Unfortunately, MIMO training and backward compatibility increases the overhead, which reduces efficiency. In environments free from legacy devices (termed *greenfield*) backward compatibility is not required.

By eliminating the components of the preamble that support backward compatibility, the greenfield format preamble is 12 μs shorter than the mixed format preamble. This difference in efficiency becomes more pronounced when the packet length is short, as in the case of VoIP traffic. Therefore, the use of the greenfield format is permitted even in the presence of legacy devices with proper MAC protection, although the overhead of the MAC protection may reduce the efficiency gained from the PHY. Range was considered as a performance metric in the PAR and comparison criteria. To increase the data rate at a given range requires enhanced robustness of the wireless link. 802.11n defines implicit and explicit TxBF (Transmit BeamForming) methods and STBC (Space-Time Block Coding), which improves link performance over MIMO with basic SDM (Spatial-Division Multiplexing). The standard also defines a new optional LDPC (Low Density Parity Check) encoding scheme, which provides better coding performance over the basic convolutional code. To break the 100 Mb/s throughput barrier, frame aggregation was added to the 802.11n MAC as the key method of increasing efficiency. The issue is that as the data rate increases, the time on air of the data portion of the packet decreases. However, the PHY and MAC overhead remain constant. This results in diminishing returns from the increase in PHY data rate. Frame aggregation increases the length of the data portion of the packet to increase overall efficiency. Two forms of aggregation exist in the standard: A-MPDU and A-MSDU. Logically, A-MSDU resides at the top of the MAC and aggregates multiple MSDUs into a single MPDU. Each MSDU is prepended with a subframe header consisting of the destination address, source address, and a length field giving the length of the SDU in bytes. This is then padded with 0 to 3 bytes to round the subframe to a 32-bit word boundary. Multiple such subframes are concatenated together to form a single MPDU. An advantage of A-MSDU is that it can be implemented in software. A-MPDU resides at the bottom of the MAC and aggregates multiple MPDUs. Each MPDU is prepended with a header consisting of a length field, 8-bit CRC, and 8-bit signature field. These subframes are similarly padded to 32-bit word boundaries. Each subframe is concatenated together. An advantage of A-MPDU is that if an individual MPDU is corrupt, the receiver can scan forward to the next MPDU by detecting the signature field in the header of the next MPDU. With A-MSDU, any bit error causes all the aggregates to fail.

| Standards | 802.11a/g | 802.11n | |
|---|---|---|---|
| | | Mandatory | Optional |
| Maximum transmission rate (Mbps) | 54 | 130 | 600 |
| Bandwidth(MHz) | 20 | 20 | 40 |
| FFT size | 64 | 64 | 128 |
| Number of subcarrier (data + pilot) | 52 (48+4) | 56 (52+4) | 114 (108+6) |
| Multi-antenna scheme | signal antenna | 2 Tx MIMO | 3,4 Tx MIMO Tx Beam forming STBC |
| Channel coding | Convolutional code (1/2, 2/3, 3/4) | Convolutional code (1/2, 2/3, 3/4, 5/6) | LDPC (1/2, 2/3, 3/4, 5/6) |
| Modulation | BPSK, QPSK, 16-QAM, 64-QAM | | |
| Spatial stream | 1 | 1 ~ 2 | 1 ~ 4 |
| Guard interval(ns) | 800 | 800 | 400 |
| Subcarrier interval | 312.5 KHz | 312.5 KHz | 312.5 KHz |
| FFT period | 3.2 μs | 3.2 μs | 3.2 μs |
| Symbol period | 4 μs | 4 μs | 4 μs |

Table 7. IEEE 802.11n OFDM parameter compared to IEEE 802.11a/g

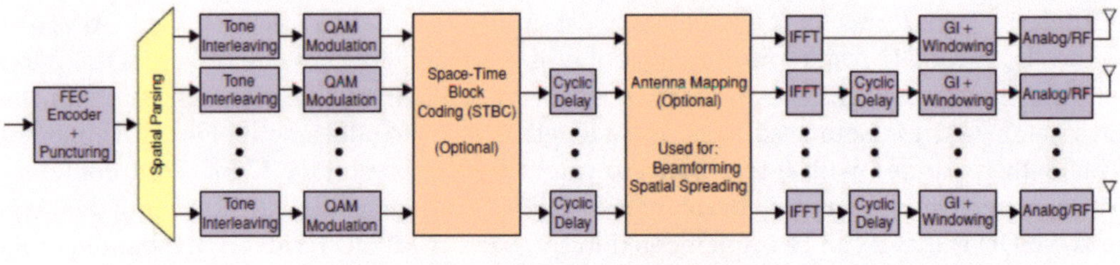


(a) General MIMO TX Datapath

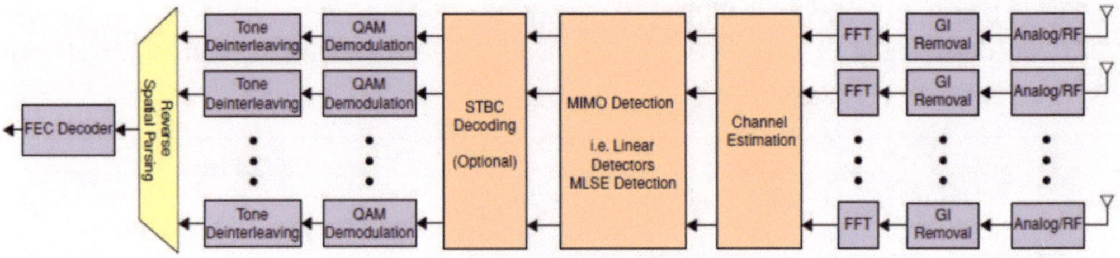


(b) General MIMO RX Datapath

Fig. 7. Block diagrams of general MIMO transmit and receive datapath structures for an IEEE 802.11n PHY [15]

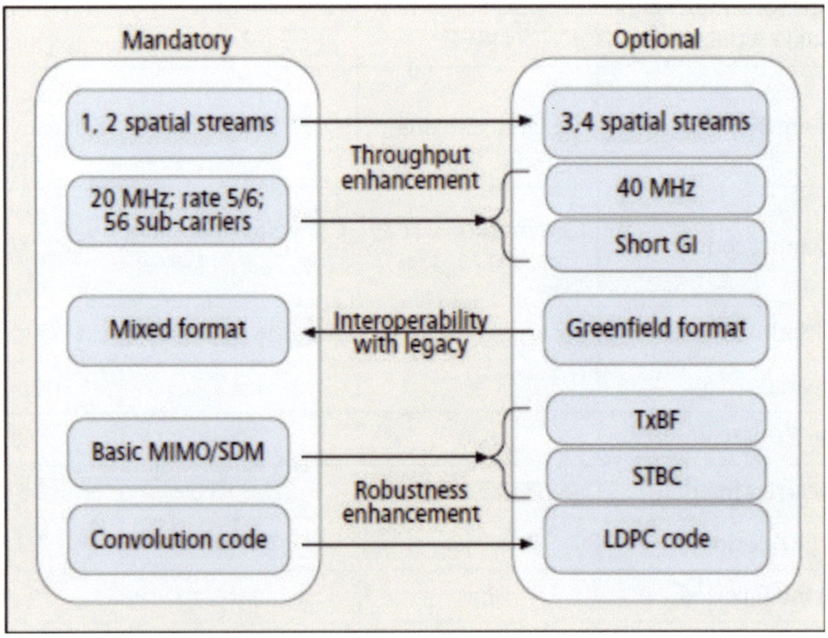


Fig. 8. Mandatory and optional 802.11n PHY features [4]

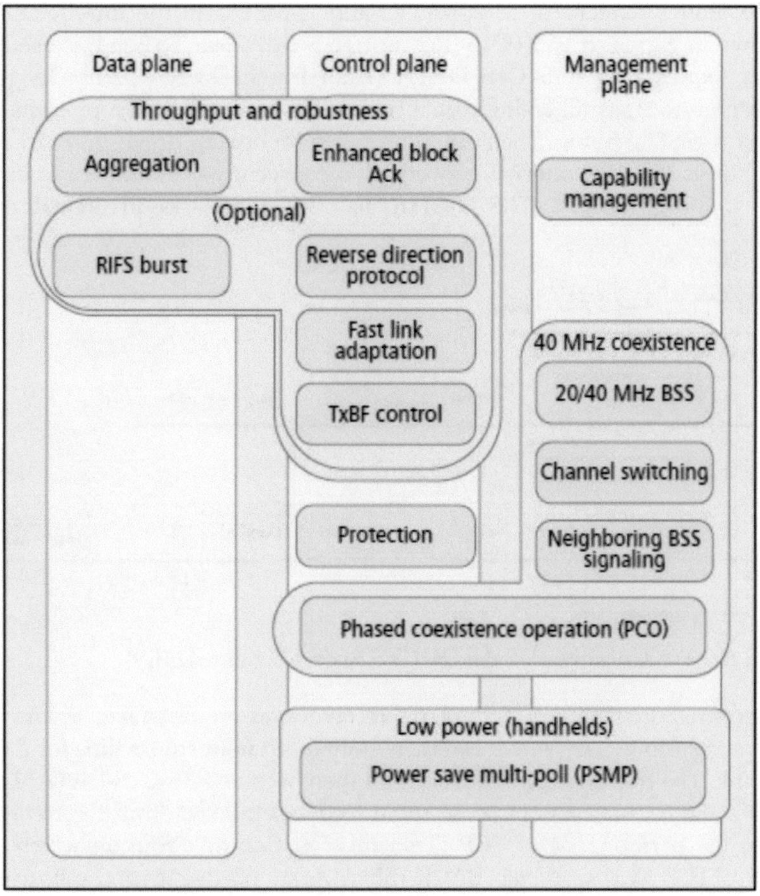


Fig. 9. Summary of 802.11n MAC enhancements [4]

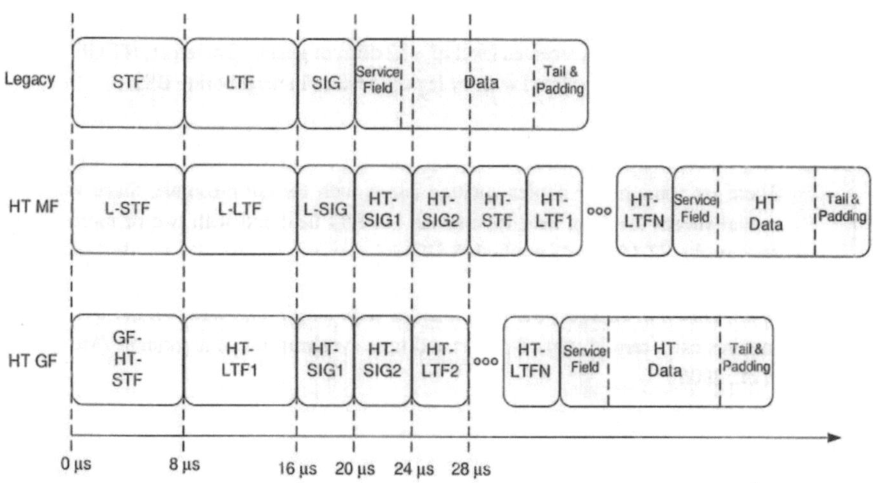


Fig. 10. Timing of the preamble fields in legacy, MF and GF in IEEE 802.11n wireless LAN[14]

The possible timing sequences for A-MPDU and A-MSDU in the uni-directional transfer case are shown in Fig. 11. If RTS/CTS (Request To Send/Clear To Send) is used, the current transmission sequence of RTS–CTS–DATA (Data frame)–ACK (Acknowledgement) only allows the sender to transmit a single data frame. The DATA frame represents either an A-MPDU or an A-MSDU frame. The system time can be broken down into virtual time slots where each slot is the time interval between two consecutive countdown of backoff timers by non-transmitting stations. The 802.11n also specifies a bi-directional data transfer method.

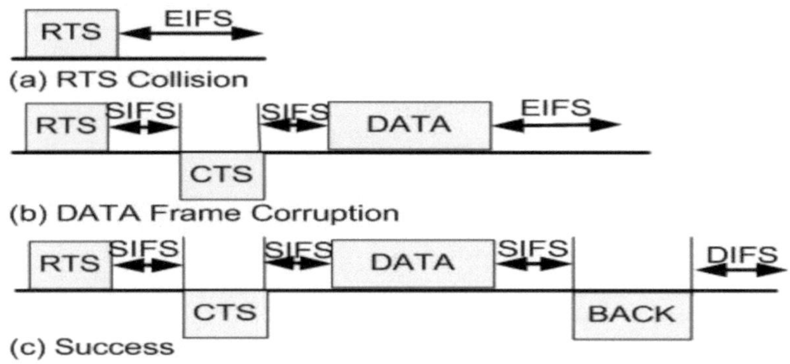


Fig. 11. IEEE 802.11n Uni-directional RTS/CTS Access Scheme [14][9]

In the bi-directional data transfer method, the receiver may request a *reverse* data transmission in the CTS control frame. The sender can then grant a certain medium time for the receiver on the reverse link. The transmission sequence will then become RTS-CTS-DATAf-DATAr-ACK. This facilitates the transmission of some small feedback packets from the receiver and may also enhance the performance of TCP (Transmission Control Protocol) which requires the transmission of TCP ACK segments. BACK (Block Acknowledgement) can be used to replace the previous ACK frame. The BACK can use a bit map to efficiently acknowledge each individual sub-frame within the aggregated frame. For the bi-directional data transfer, the reverse DATAr frame can contain a BACK to acknowledge the previous DATAf frame. In this subsection, we briefly mention the most important MAC enhancements with a more detailed explanation on frame aggregation, which maximizes throughput and efficiency. Aggregate exchange sequences are made possible with a protocol that acknowledges multiple MPDUs with a single block ACK in response to a block acknowledgment request (BAR). Another key enhancement that the 802.11n specifies is the bidirectional data transfer method over a single TXOP, known as reverse direction. This feature permits the transportation of data frames, even aggregates, in both directions in one TXOP. Until now, when the sender STA is allocated with a TXOP, it informs surrounding STAs about how long the wireless medium will be engaged. However, this approximation of channel use is not always accurate, and often the transmission ends sooner. As a result, contended STAs assume that the channel is still occupied when this is not the case. With reverse direction, the initial receiver STA is allowed to send any packets available that are addressed to the sender for the remaining TXOP time. This fits especially well with TCP because it allows a TCP link to *piggyback* TCP ACK collection onto TCP data transmission. The long-NAV (Network Allocation Vector) is another enhancement that improves scheduling, given that a station that holds a TXOP may set a longer NAV value intended to protect multiple PPDUs. Another mandatory feature is PCO (Phased Coexistence

Operation) which protects stations using either 20 MHz or 40 MHz channel spectrum at the same time. Finally, the RIFS (Reduced IFS) is proposed to allow a time interval of 2 µs between multiple PPDUs, which is much shorter than SIFS as defined in the legacy standards.

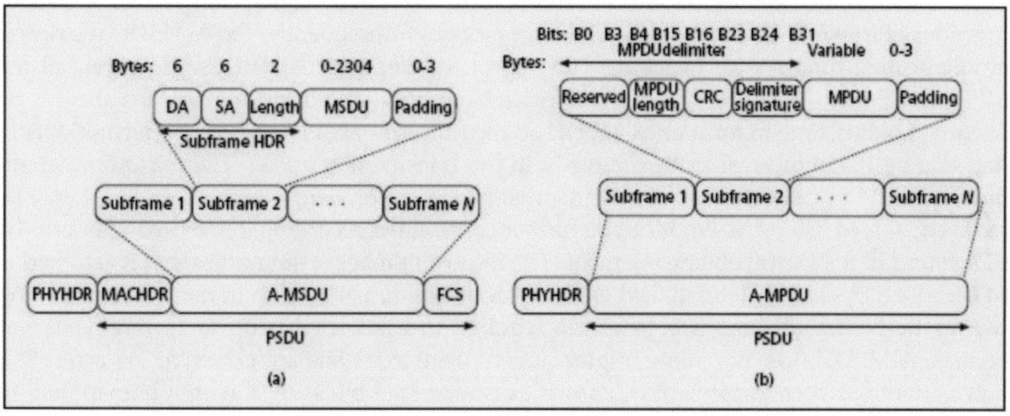


Fig. 12. One-level frame aggregation: a) A-MSDU; b) A-MPDU [10]

**A-MSDU** – The principle of the A-MSDU (or MSDU aggregation) is to allow multiple MSDUs to be sent to the same receiver concatenated in a single MPDU. This definitely improves the efficiency of the MAC layer, specifically when there are many small MSDUs, such as TCP acknowledgments. This supporting function for A-MSDU within the 802.11n is mandatory at the receiver. For an A-MSDU to be formed, a layer at the top of the MAC receives and buffers multiple packets (MSDUs). The A-MSDU is completed either when the size of the waiting packets reaches the maximal A-MSDU threshold or the maximal delay of the oldest packet reaches a pre-assigned value. Its maximum length can be either 3839 or 7935 bytes; this is 256 bytes shorter than the maximum PHY PSDU length (4095 or 8191 bytes, respectively), as predicted space is allocated for future status or control information. The size can be found in the HT capabilities element that is advertised from an HT STA in order to declare its HT status. The maximal delay can be set to an independent value for every AC but is usually set to 1 µs for all ACs. There are also certain constraints when constructing an A-MSDU:

- All MSDUs must have the same TID value
- Lifetime of the A-MSDU should correspond to the maximum lifetime of its constituent elements
- The DA (Destination Address) and SA (Sender Address) parameter values in the subframe header must match to the same RA (Receiver Address) and TA (Transmitter Address) in the MAC header.

Thus, broadcasting or multicasting is not allowed. Figure 12a describes a simple structure of a carrier MPDU that contains an A-MSDU. Each subframe consists of a subframe header followed by the packet that arrived from the LLC and 0 ~ 3 bytes of padding. The padding size depends on the rule that each subframe, except for the last one, should be a multiple of four bytes, so the end receiver can approximate the beginning of the next subframe. A major drawback of using A-MSDU is under error-prone channels. By compressing all MSDUs into a single MPDU with a single sequence number, for any subframes that are corrupted, the entire A-MSDU must be retransmitted. Additional frame structures or optimum frame sizes have been proposed to improve performance under noisy channels.

**A-MPDU** – The concept of A-MPDU aggregation is to join multiple MPDU subframes with a single leading PHY header. A key difference from A-MSDU aggregation is that A-MPDU functions after the MAC header encapsulation process. Consequently, the A-MSDU restriction of aggregating frames with matching TIDs is not a factor with A-MPDUs. However, all the MPDUs within an A-MPDU must be addressed to the same receiver address. Also, there is no waiting/holding time to form an A-MPDU so the number of MPDUs to be aggregated totally depends on the number of packets already in the transmission queue. The maximum length that an A-MPDU can obtain – in other words the maximum length of the PSDU that may be received – is 65,535 bytes, but it can be further constrained according to the capabilities of the STA found in the HT capabilities element. The utmost number of subframes that it can hold is 64 because a block ACK bitmap field is 128 bytes in length, where each frame is mapped using two bytes. Note that these two bytes are required to acknowledge up to 16 fragments but because A-MPDU does not allow fragmentation, these extra bits are excessive. As a result, a new variant has been implemented, known as compressed block ACK with a bitmap field of eight bytes long. Finally, the size of each subframe is limited to 4095 bytes as the length of a PPDU cannot exceed the 5.46-ms time limit; this can be derived from the maximum length divided by the lowest PHY rate, which is 6 Mb/s and is the highest duration of an MPDU in 802.11a. The basic structure is shown in Fig. 12b. A set of fields, known as delimiters are inserted before each MPDU and padding bits varied from 0 ~ 3 bytes are added at the tail. The basic operation of the delimiter header is to define the MPDU position and length inside the aggregated frame. It is noted that the CRC (Cyclic Redundancy Check) field in the delimiter verifies the authenticity of the 16 preceding bits. The padding bytes are added such that each MPDU is a multiple of four bytes in length, which can assist subframe delineation at the receiver side. In other words, the MPDU delimiters and PAD bytes determine the structure of the A-MPDU. After the AMPDU is received, a de-aggregation process initiates. First it checks the MPDU delimiter for any errors based on the CRC value. If it is correct, the MPDU is extracted, and it continues with the next subframe till it reaches the end of the PSDU. Otherwise, it checks every four bytes until it locates a valid delimiter or the end of the PSDU. The delimiter signature has a unique pattern to assist the de-aggregation process while scanning for delimiters.

### 4.3 IEEE 802.11ac/ad PHY/MAC layer

The WiGig (Wireless Gigabit) Alliance was formed to meet this need by establishing a unified specification for wireless communication at multi-gigabit speeds; this specification is designed to drive a global ecosystem of interoperable products. The WiGig MAC and PHY Specification enables data rates up to 7 Gbps, more than 10 times the speed of the fastest Wi-Fi networks based on IEEE 802.11n. It operates in the unlicensed 60 GHz frequency band, which has much more spectrum available than the 2.4 GHz and 5 GHz bands used by existing Wi-Fi products. This allows wider channels that support faster transmission speeds. The WiGig specification is based on the existing IEEE 802.11 standard, which is at the core of hundreds of millions of Wi-Fi products deployed worldwide. The specification includes native support for Wi-Fi over 60 GHz; new devices with tri-band radios will be able to seamlessly integrate into existing 2.4 GHz and 5 GHz Wi-Fi networks. The specification enables a broad range of advanced uses, including wireless docking and connection to displays, as well as virtually instantaneous wireless backups, synchronization and file transfers between computers and handheld devices. For the first time, consumers will be able to create a complete computing and consumer electronics experience without wires. The WiGig specification includes key features to maximize performance, minimize implementation complexity and cost, enable compatibility with existing Wi-Fi and provide advanced security. Key features include:

- Support for data transmission rates up to 7 Gbps; all devices based on the WiGig specification will be capable of gigabit data transfer rates
- Designed from the ground up to support low-power handheld devices such as cell phones, as well as high-performance devices such as computers; includes advanced power management
- Based on IEEE 802.11; provides native Wi-Fi support and enables devices to transparently switch between 802.11 networks operating in any frequency band including 2.4 GHz, 5 GHz and 60 GHz
- Support for beamforming, maximizing signal strength and enabling robust communication at distances beyond 10 meters
- Advanced security using the Galois/Counter Mode of the AES encryption algorithm
- Support for high-performance wireless implementations of HDMI, DisplayPort, USB and PCIe

The WiGig specification defines PHY and MAC layers and is based on IEEE 802.11. This enables native support for IP networking over 60 GHz. It also makes it simpler and less expensive to produce devices that can communicate over both WiGig and existing Wi-Fi using tri-band radios (2.4 GHz,5 GHz and 60 GHz).

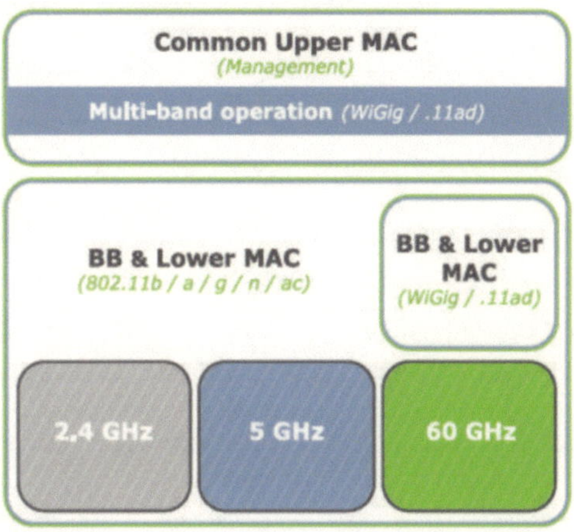


Fig. 13. WiGig architecture enables tri-band communications [16]

The WiGig Alliance is also defining PALs (Protocol Adaptation Layers) that support specific data and display standards over 60 GHz. PALs allow wireless implementations of these standard interfaces that run directly on the WiGig MAC and PHY, as shown in Figure 14, and can be implemented in hardware. The initial PALs are audio-visual (A/V), which defines support for HDMI and DisplayPort, and input-output (I/O), which defines support for USB and PCIe.

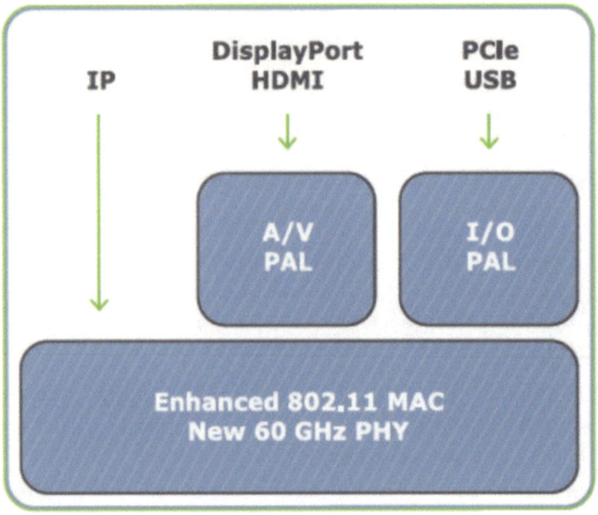


Fig. 14. WiGig Protocol Adaptation Layers (PALs) [16]

| MCS | Modulation | R | $N_{BPSC}$ | $N_{SD}$ | $N_{SP}$ | $N_{CBPS}$ | $N_{DBPS}$ | $N_{SS}$ | Data rate [Mbps] | |
|---|---|---|---|---|---|---|---|---|---|---|
| | | | | | | | | | 800 ns GI | 400ns GI |
| 1 | 64-QAM | 2/3 | 6 | 228 | 8 | 5,472 | 3,648 | 4 | 912 | 1,013 |
| 2 | 64-QAM | 3/4 | 6 | 228 | 8 | 5,472 | 4,104 | 4 | 1,026 | 1,140 |
| 3 | 64-QAM | 5/6 | 6 | 228 | 8 | 5,472 | 4,560 | 4 | 1,140 | 1,266 |

Table 8. Transmission rate performance of IEEE 802.11ac [17]

| Modulation with RS(255,233) | Data rate [Mbps] | |
|---|---|---|
| | Narrowband mode | Wideband mode |
| BPSK-1/2 | 162 | 708.6 |
| BPSK-3/4 | 243 | 1062.9 |
| QPSK-1/2 | 323 | 1417.2 |
| QPSK-3/4 | 486 | 2125.7 |
| 16-QAM-1/2 | 647.8 | 2834.3 |
| 16-QAM-3/4 | 971.75 | 4251.4 |
| 64-QAM-2/3 | 1295.66 | 5668.5 |

Table 9. Transmission rate performance of IEEE 802.11ad [18]

| Parameter | Value |
|---|---|
| FFT size | 256 |
| Number of total sub-carrier | 244 |
| Data subcarrier | 228 |
| Sub-carrier spacing | 312.5 KHz |
| FFT period | 3.2 μs |
| Guard interval | 800 ns, 400 ns, 200 ns |
| Symbol period | 4 μs |

Table 10. IEEE 802.11ac OFDM parameter [17]

| Parameter | Value | |
|---|---|---|
| | Narrowband mode | Wideband mode |
| Channel bandwidth | 540/720 MHz | 2,160 MHz |
| FFT bandwidth | 576 MHz | 2,304 MHz |
| FFT size | 256 | 1,024 |
| Sub-carrier spacing | 2.25 MHz | 2.25 MHz |
| Guard interval | 111 ns | 111ns |
| FFT period | 444 ns | 444 ns |
| OFDM symbol time | 556 ns | 556 ns |
| Data sub-carriers | 192 | 768 |
| Pilot/Zero sub-carriers | 16/5 | 60/5 |
| Nominal used bandwidth | 479.25 MHz | 2,049.75 MHz |

Table 11. IEEE 802.11ad OFDM parameter [18]

## 5. Frame error rate and DCF throughput analysis

### 5.1 Frame error rate

#### 5.1.1 Frame error rate of fixed wireless channel

In IEEE 802.11a/g wireless LAN, fixed wireless channel is assumed to be Rayleigh fading channel. The probability of bit error is upper bound by

$$P_b < \frac{1}{k} \sum_{d=d_{free}}^{\infty} B_d P_d \tag{1}$$

where $d_{free}$ is the free distance of the convolutional code, $B_d$ is the total number of information bit ones on all weight $d$ paths, $P_d$ is the probability of selecting a weight $d$ output sequence as the transmitted code sequence, and $k$ is the number of information bits per clock cycle. Because the weight structure is generally accepted that the first five terms in equation (1) dominate, equation (1) can be rewritten as

$$P_b < \frac{1}{k} \sum_{d=d_{free}}^{d_{free}+4} B_d P_d \tag{2}$$

The probability of selecting the incorrect path when $d$ is odd.

$$P_d = \sum_{i=\frac{d+1}{2}}^{d} \binom{d}{i} p^i (1-p)^{d-i} \tag{3}$$

where $p$ is the probability of channel bit error. The probability of selecting the incorrect path when $d$ is even.

$$P_d = \sum_{i=\frac{d}{2}+1}^{d} \binom{d}{i} p^i (1-p)^{d-i} + \frac{1}{2} \binom{d}{d/2} p^{d/2} (1-p)^{d/2} \tag{4}$$

To achieve data rates of 54 Mbps for wireless access, the IEEE 802.11a standard utilizes MQAM($q=6$, $M=64$) with convolutional coding at rate $r = 3/4$. We obtain the approximate channel bit error probability for the $i^{th}$ sub-channel for MQAM with a square constellation as [19]

$$p_i \cdot \approx \frac{4\left(1-\frac{1}{\sqrt{M}}\right) \cdot e^{-d\cdot\zeta_i\left[\frac{3qr\overline{\gamma_{b_i}}}{3qr\overline{\gamma_{b_i}}+2(M-1)(\zeta_I+1)}\right]}}{q\sqrt{2\pi c_2}\left[\frac{3qr\overline{\gamma_{b_i}}+2(M-1)(\zeta_I+1)}{2(M-1)(\zeta_I+1)}\right]^d} - \frac{2\left(1-\frac{1}{\sqrt{M}}\right)^2 \cdot e^{-d\cdot\zeta\left[\frac{3qr\overline{\gamma_{b_i}}}{3qr\overline{\gamma_{b_i}}+2(M-1)(\zeta_I+1)}\right]}}{\pi c_2 q\left[\frac{3qr\overline{\gamma_{b_i}}+(M-1)(\zeta_I+1)}{(M-1)(\zeta_I+1)}\right]} \tag{5}$$

where $c_2 = 2.6 + 0.1\zeta$ is empirically obtained and d=1 for HDD. $\zeta_i$ is the ratio of direct-to-diffuse signal power on the $i^{th}$ sub-channel. $\zeta$ has 0 in a pure Rayleigh fading channel and ranges from 0 to 10 in a composite Rayleigh/Ricean fading channel. $\overline{\gamma_{bi}}$ is the ratio of received average energy per bit-to-noise power spectral density on the $i^{th}$ sub-channel. The overall $p$ is the average of the probability of bit error on each of the $N$ OFDM sub-channels.

$$p = \frac{1}{N}\sum_{i=1}^{N} p_i \tag{6}$$

Note that for either no channel fading or for all sub-channels experiencing the same fading (that is, $\zeta_i = \zeta$ and $\overline{\gamma_{b_i}} = \overline{\gamma_b}$ for all $i$ ), then $p_i = p$ . $\overline{\gamma_b} = \overline{E_b} / N_o$ is the ratio of received average energy per bit-to-noise power spectral density , $\zeta$ is the ratio of direct-to-diffuse signal power. Now, using equation (6) in equation (3) or (4) and taking the result into equation (2), we obtain the performance of 64 QAM with HDD over Ricean fading channels. For basic access mechanism, a data packet including the PHY header and the MAC header needs retransmission if any one bit of them is corrupted. We define a variable $P_C$ which is the probability that a backoff occurs in a station due to bit errors in frames. We further assume that bit errors randomly appear in the frames. So frame error rate is represented by (7).

$$P_C = 1 - (1 - P_b)^{L_{preamble} + PHY_h + MAC_h + P + L_{ACK}} \tag{7}$$

CSMA/CA is also used as the MAC scheme in IEEE 802.11n wireless LAN, and it has basic and RTS/CTS access scheme. Although there is a successful RTS/CTS transmission in the time slot, a frame have to be retransmitted when there is a bit error in a payload. For convenience, we define a variable $P_e$ which is the probability that a backoff occurs in a station due to bit errors in frames. We further assume that bit errors randomly appear in the frames and A-MSDU scheme is used. So frame error rate is represented by (8).

$$P_e = 1 - (1 - q)^L \tag{8}$$

Where $L$ is the aggregated MAC frame's size. For a convolutional code with a coding rate $k_c/n_c$, the bit error rate, denoted as $q$, can be approximated by

$$q = \frac{1}{k_c}\sum_{i=\frac{d_{free}+1}{2}}^{d_{free}} \binom{d_{free}}{i} (q_b)^i (1-q_b)^{d_{free}-i} \quad (d_{free} \text{ is odd}) \tag{9}$$

$$q = \frac{1}{k_c}\sum_{i=\frac{d_{free}}{2}+1}^{d_{free}} \binom{d_{free}}{i} (q_b)^i (1-q_b)^{d_{free}-i} + \frac{1}{2k_c}\binom{d_{free}}{d_{free}/2} (q_b)^{d_{free}/2} (1-q_b)^{d_{free}/2} \quad (d_{free} \text{ is even})$$

Where $d_{free}$ is the maximum free distance of the convolutional code and $q_b$ is the probability of a bit error for the M-QAM.

$$q_b = \frac{Mq_s}{2(M-1)} \tag{10}$$

$q_s$ is the SER(Symbol Error Rate) under the Rician fading channel.

$$q_s = \left(\frac{1+K}{1+K+\frac{\rho(d_{\min})^2}{8}}\right)^4 e^{-4\left(\frac{\frac{\rho(d_{\min})^2}{8}(\| H \|)^2}{1+K+\frac{\rho(d_{\min})^2}{8}}\right)} \tag{11}$$

$K$ is the Rician factor and ρ may be interpreted as the average SNR at the receive antenna in a SISO fading link. $d_{\min}$ is the minimum distance of separation of the underlying scalar constellation. $H$ is $M_R \times M_T$ channel transfer function and $\| H \|^2$ is the squared Frobenius norm of the channel.

### 5.1.2 Frame error rate of mobile wireless channel

Mobile wireless channel is assumed to be flat fading Rayleigh channel with Jake spectrum. The channel is in fading states or inter-fading states by evaluating a certain threshold value of received signal power level. If and only if the whole frame is in inter-fading state, there is the successful frame transmission. If any part of frame is in fading duration, the frame is received in error. In the fading channel fading margin is considered and defined as $\rho = R_{req}/R_{rms}$, Where $R_{req}$ is the required received power level and $R_{rms}$ is the mean received power. Generally, the fading duration and inter-fading duration can be taken to be exponentially distributed for ρ<-10dB. With the above assumptions, let $Tpi$ be the frame duration, then the frame error rate is given by (12).

$$FER = 1 - \frac{Ti}{Ti + T_f} P(t_i > Tpi) \tag{12}$$

Where, $t_i$ is inter-fading duration and $t_f$ is fading duration. $Ti$ is the mean value of the random variable $t_i$ and $T_f$ is the mean value of the random variable $t_f$. $P(ti > Tpi)$ is the probability that inter-fading duration lasts longer than $Tpi$. Since exponential distribution is assumed for $t_i$, $P(t_i > Tpi) = \exp(-\frac{Tpi}{Ti})$. For Rayleigh fading channel, the average fading duration is given by (13).

$$Ti = \frac{\exp(\rho) - 1}{fd\sqrt{2\pi\rho}} \tag{13}$$

$T_i + T_f$ is $\frac{1}{N_f}$, where $N_f$ is the level crossing rate, which is given by $fd\sqrt{2\pi\rho}\exp(-\rho)$. $f_d$ is the maximum Doppler frequency and evaluated as $\frac{v}{\lambda}$. $v$ is the mobile speed and $\lambda$ is wavelength. Frame error rate can be expressed by (14) [11].

$$FER = 1 - \exp(-\rho - f_d\sqrt{2\pi\rho}Tpi) \quad (14)$$

Equation (14) shows that frame error rate is determined by fading margin, maximum Doppler frequency and frame duration. Since fading margin and maximum Doppler frequency are hard to dynamically control, the only controllable parameter is frame duration to get required frame error rate. For the RTS/CTS access mode, the frame duration $T_{pi}$ is $T_H + T_{RTS} + T_{CTS} + T_{DATA} + T_{ACK}$. $T_H$ is preamble transmission time + PLCP header transmission time + MAC header transmission time. $T_{DATA}$ is MSDU transmission time and $T_{ACK}$ is ACK frame transmission time. $T_{RTS}$ is RTS frame transmission time and $T_{CTS}$ is CTS frame transmission time.

### 5.1.3 Numerical results

#### 5.1.3.1 Analysis of frame error rate under the Rayleigh/Rician fading channel with fixed stations

In the Fig. 15, $P_c(P, \gamma_b, K)$ shows PER(Packet Error Rate) due to $\gamma_b$, the ratio of received average energy per bit-to-noise power spectral density. K means Rician factor and P means payload size. and as expected, the PER (Frame Error Rate) performance improves with K and the smaller payload size is, the better performance is.

In the Fig. 16, $q_s(\rho,K)$ shows SER(Symbol Error Rate) and $P_e(K,\rho,n_s,P)$ shows PER(Packet Error Rate). K means Rician factor and as expected, the SER performance improves with K. Also, the PER performance improves with K and the smaller subframe' payload size is, the better performance is.

#### 5.1.3.2 Analysis of frame error rate under the flat fading Rayleigh channel with mobile stations

In the Fig. 17(a) ~ Fig. 17(c), the symbol fer ($\rho$, $v$, P) shows frame error rate of IEEE 802.11a/g. In the Fig. 17(d), the symbol fer (ns, $\rho$, $v$, P) shows frame error rate of IEEE 802.11n with the horizontal parameter of subframe' payload size. In the Fig. 17(e), the symbol fer ($\rho$, ns, $v$, P) shows frame error rate of IEEE 802.11n using the number of subframes as the horizontal parameter. It is generally identified that the higher mobile speed is, the higher frame error rate is. In case of payload size, the same result mentioned above is also acquired.

### 5.2 DCF throughput analysis

The back-off procedure of the DCF protocol is modeled as a discrete-time, two-dimensional Markov chain. Fig. 18 shows the Bianchi's Markov chain model for the back-off window size. We define $W = CW_{\min}$. Let $m$, the maximum back-off stage, be such value that $CW_{\max} = 2^m W$. We also define $W_i = 2^i W$, where $i \in (0,m)$ is called the back-off stage. Let $s(t)$ be the stochastic process representing the back-off stage $(0,\ldots,m)$ of the station at time $t$. $p$ is the probability that a transmission is collided or unsuccessfully executed.

We will present the analytical evaluation of saturation throughput with bit errors appearing in the transmitting channel. The number of stations $n$ is assumed to be fixed and each station always has packets for transmission. In other words, we operate in saturation conditions, the transmission queue of each station is assumed to be always nonempty.

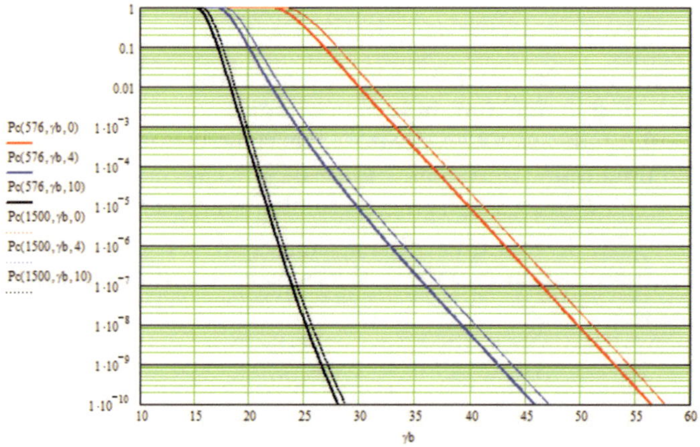


(a) 802.11a OFDM

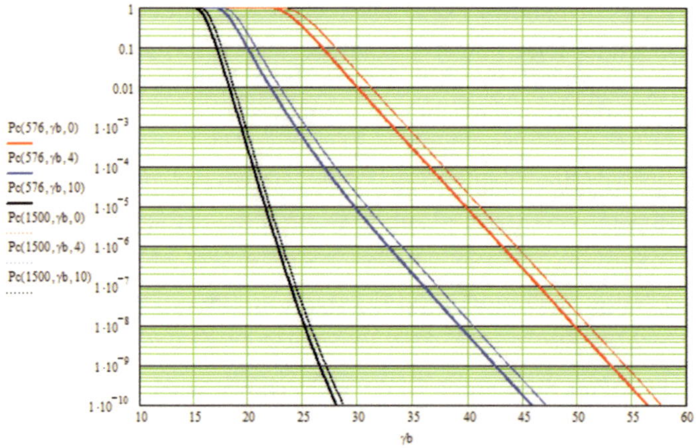


(b) 802.11g ERP-OFDM

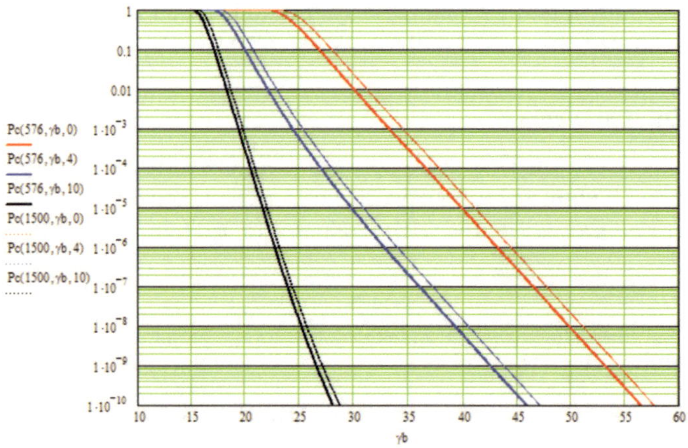


(c) 802.11g DSSS-OFDM

Fig. 15. Frame error rate of IEEE 802.11a/g fixed LAN over Rayleigh fading channel

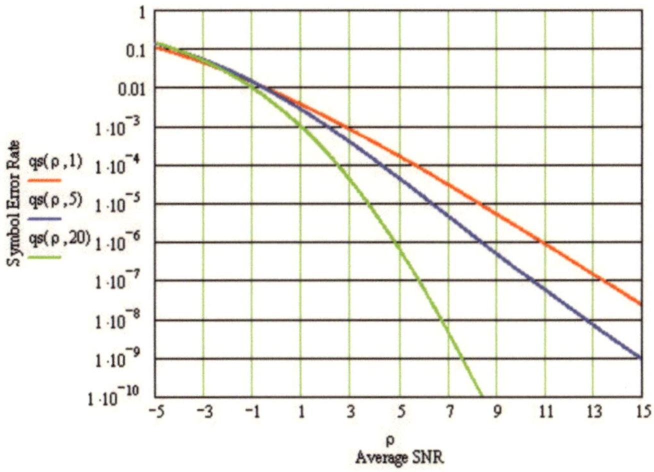


(a) Symbol error rate

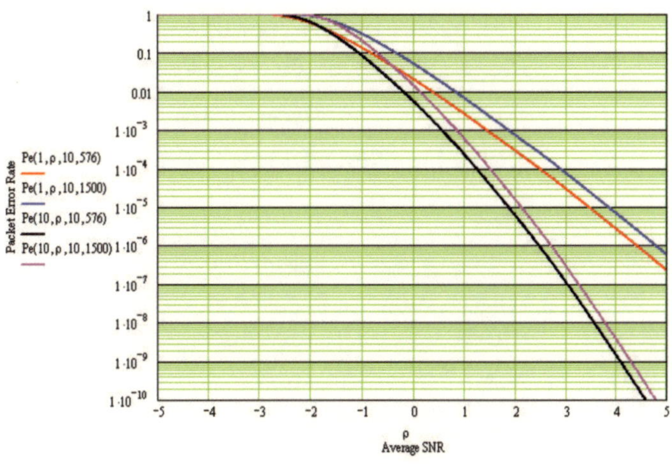


(b) Packet error rate

Fig. 16. Frame error rate of IEEE 802.11n fixed LAN over Rician fading channel

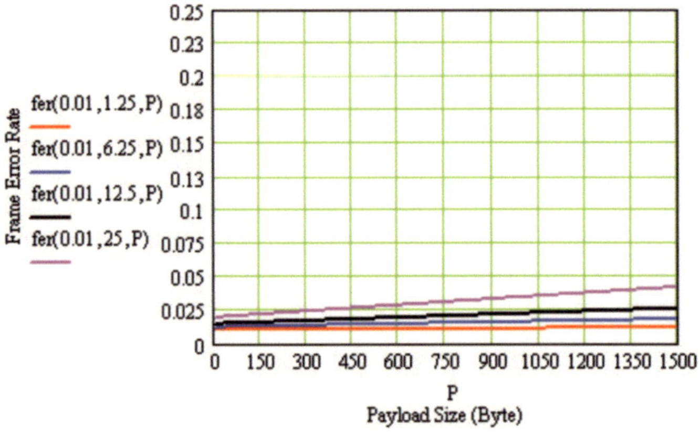


(a) 802.11a OFDM (54 Mbps)

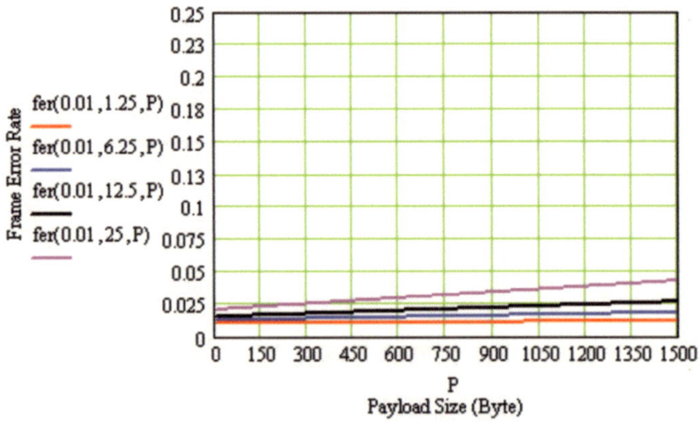


(b) 802.11g ERP-OFDM (54 Mbps)

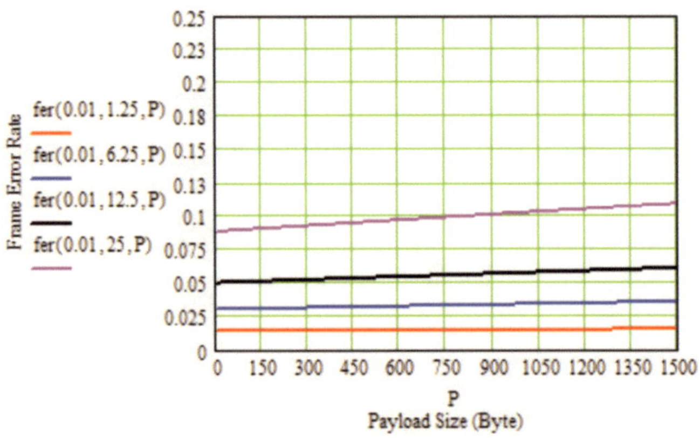


(c) 802.11g DSSS-OFDM (54 Mbps)

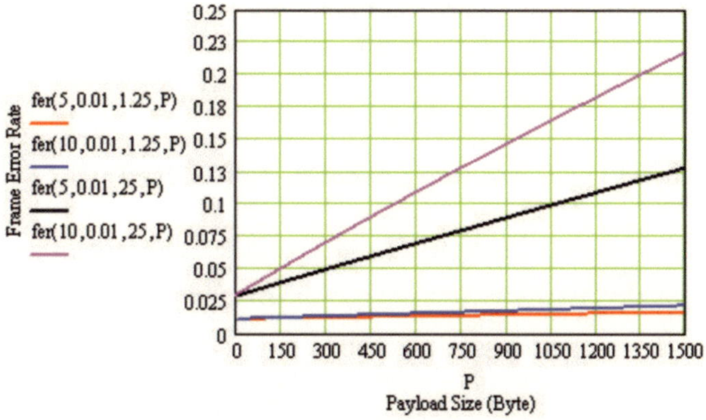


(d) 802.11n OFDM (58.5 Mbps, Payload size)

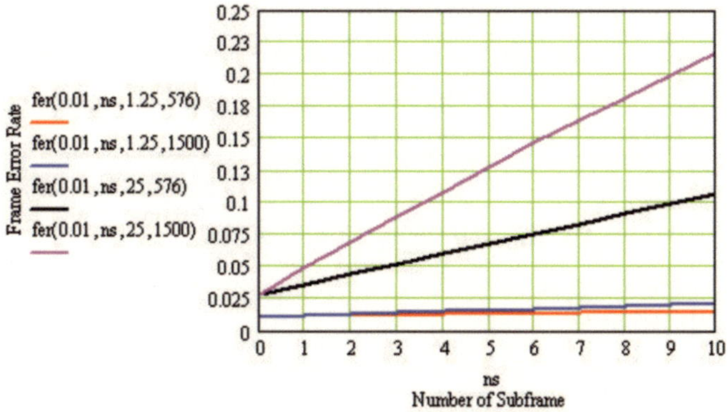


(e) 802.11n OFDM (58.5 Mbps, number of subframe)

Fig. 17. Frame error rate of IEEE 802.11a/g/n mobile LAN

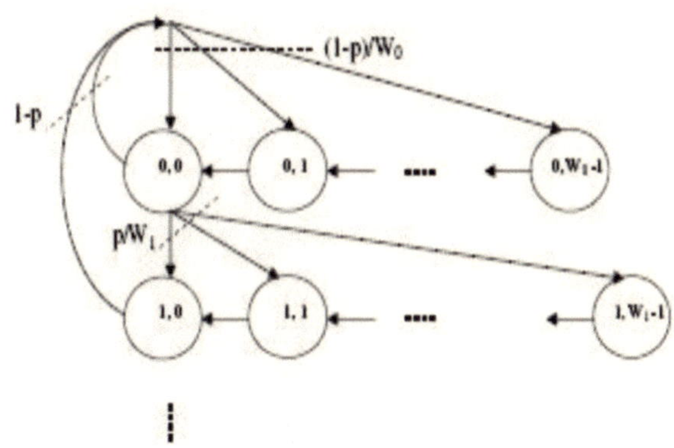


Fig. 18. Markov chain model for the backoff window size [20]

Let $S$ be the normalized system throughput, defined as the fraction of time in which the channel is used to successfully transmit payload bits. $P_{tr}$ is the probability that there is at least one transmission in the considered slot time. Since $n$ stations contend on the channel and each transmits with probability $\tau$, we get

$$P_{tr} = 1-(1-\tau)^n \tag{15}$$

Table 12 shows physical and MAC layer parameters of IEEE 802.11a/g/n-based wireless LAN [21].

| Parameter | Explanation |
|---|---|
| FER | Frame error rate |
| $\tau$ | Packet transmission probability |
| $n$ | Number of stations |
| $P$ | Payload size |
| $T_{RTS}$ | RTS frame transmission time |
| $T_{CTS}$ | CTS frame transmission time |
| $T_H$ | PLCP preamble transmission time + PLCP header transmission time + MAC header transmission time |
| $T_{DATA}$ | Payload transmission time |
| $T_{ACK}$ | ACK frame transmission time |
| $T_{BACK}$ | Block ACK frame transmission time |
| $\sigma$ | Slot time |
| $T_{SIFS}$ | SIFS time |
| $T_{DIFS}$ | DIFS time |
| $T_{EIFS}$ | EIFS time |
| $CW_{\min}$ | Minimum backoff window size |
| $CW_{\max}$ | Maximum backoff window size |

Table 12. IEEE 802.11a/g/n PARAMETERS

### 5.2.1 IEEE 802.11a/g DCF throughput

Saturation throughput is represented as shown in (16)[5].

$$S = \frac{P_s P_{tr} P}{(1-P_{tr})\sigma + P_{tr}P_sT_s + P_{tr}(1-P_s)T_c} = \frac{n\tau(1-\tau)^{n-1}(1-FER)P}{(1-\tau)^n\sigma + n\tau(1-\tau)^{n-1}(1-FER)T_s + [1-(1-\tau)^n]T_c - n\tau(1-\tau)^{n-1}(1-FER)T_c} \tag{16}$$

$P_s$ is the probability that a transmission successfully occurs on the channel and is given by the probability that exactly one station transmits on the channel, conditioned on the fact that at least one station transmits.

$$P_s = \frac{n\tau(1-\tau)^{n-1}(1-FER)}{P_{tr}} \tag{17}$$

The average amount of payload information successfully transmitted in a slot time is $P_{tr}P_sP$, since a successful transmission occurs in a slot time with probability $P_{tr}P_s$. The average length of a slot time is readily obtained considering that, with probability $1-P_{tr}$, the channel is empty, with probability $P_{tr}P_s$ it contains a successful transmission, and with probability $P_{tr}(1-P_s)$ it contains a collision. Where $T_s$ is the average time the channel is sensed busy because of a successful transmission, and $T_c$ is the average time the channel is sensed busy by each station during a collision or error. $\sigma$ is the duration of an empty slot time. In the RTS/CTS access scheme, we obtain,

$$T_S = T_{RTS} + T_{CTS} + T_{DATA} + T_{ACK} + T_{DIFS} + 3T_{SIFS} \tag{18}$$

$$T_C = T_{RTS} + T_{EIFS} = T_{RTS} + T_{SIFS} + T_{ACK} + T_{DIFS}$$

### 5.2.2 IEEE 802.11n DCF throughput

The saturation throughput can be calculated as follows [9].

$$S = \frac{E_p}{E_t} = \frac{L_p P_{tr} P_s (1-P_e)}{T_{idle}P_{idle} + T_c P_{tr}(1-P_s) + T_e P_{err} + T_{succ}P_{succ}} = \frac{L_p n\tau(1-\tau)^{n-1}(1-P_e)}{(1-\tau)^n\sigma + n\tau(1-\tau)^{n-1}(1-P_e)T_{succ} + [1-(1-\tau)^n - n\tau(1-\tau)^{n-1}]T_c + n\tau(1-\tau)^{n-1}P_eT_e} \tag{19}$$

where $E_p$ is the number of payload information bits successfully transmitted in a virtual time slot, and $E_t$ is the expected length of a virtual time slot. $P_e$ is the error probability on condition that there is a successful RTS/CTS transmission in the time slot. $P_{idle}$ is the probability of an idle slot. $P_s$ is the probability for a non-collided transmission. $P_{err}$ is the transmission failure probability due to error (no collisions but having transmission errors). $P_{succ}$ is the probability for a successful transmission without collisions and transmission errors. $T_{idle}$, $T_c$ and $T_{succ}$ are the idle, collision and successful virtual time slot's length. $T_e$ is the virtual time slot length for an error transmission sequence. $L_p$ is the aggregated frame's payload length. In the RTS/CTS scheme, we obtain,

$$T_C = T_{RTS} + T_{EIFS} \tag{20}$$

$$T_{succ} = T_{RTS} + T_{CTS} + T_{DATA} + T_{BACK} + 3T_{SIFS} + T_{DIFS}$$

$$T_e = T_{RTS} + T_{CTS} + T_{DATA} + T_{EIFS} + 2T_{SIFS}$$

### 5.2.3 Numerical results

#### 5.2.3.1 Analysis of IEEE 802.11a/g/n DCF saturation throughput under the Rayleigh fading channel with fixed stations

We analyze and compare the DCF throughput with the maximum physical transmission rate of 54 Mbps between IEEE 802.11a and 802.11g-based wireless Internet as shown in Fig. 19. Of the four different physical layers defined in the IEEE 802.11g standard, ERP-OFDM and DSSS-

OFDM are used. Three common packet sizes of 60 bytes(TCP ACK), 576 bytes(typical size for web browsing) and 1,500 bytes(the maximum size for Ethernet) are considered.

In the Fig.19, S( $P$ , $\gamma_b$ , $\zeta$ , $n$ , $\tau$ ) shows the saturation throughput over error-prone channel due to number of stations( $n$ ) for common packet sizes on the condition that packet transmission probability( $\tau$ ), average energy per bit-to-noise power spectral density( $\gamma_b$ ) and the ratio of direct-to-diffuse signal power( $\zeta$ ) are fixed.

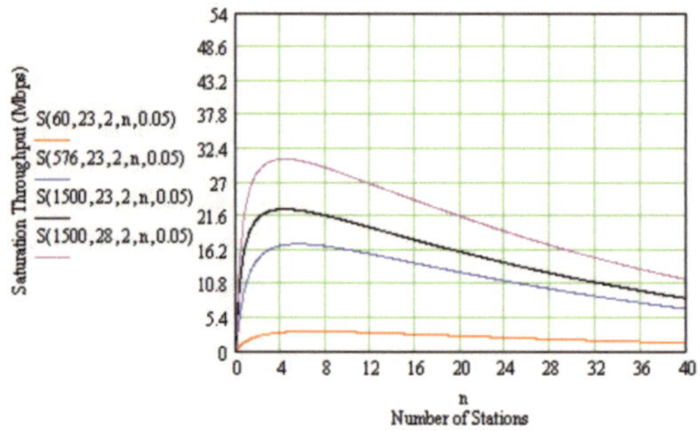


(a) IEEE 802.11a

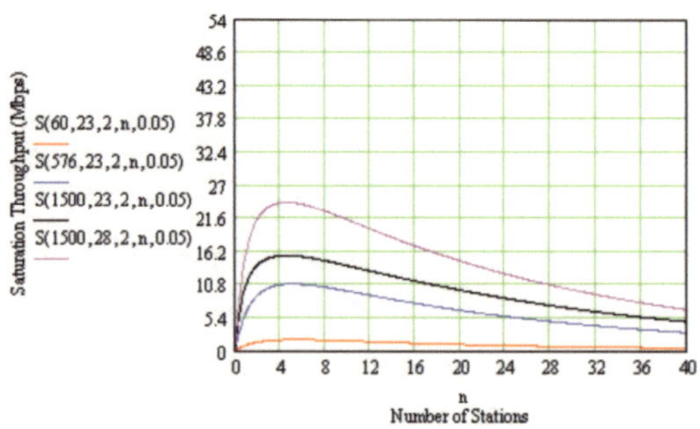


(b) IEEE 802.11g ERP-OFDM

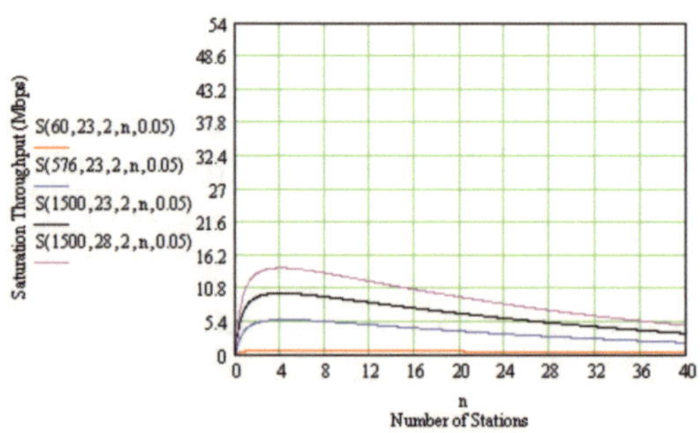


(c) IEEE 802.11g DSSS-OFDM

Fig. 19. DCF throughput in 802.11a/g ERP-OFDM/g DSSS-OFDM (54 Mbps)

The larger payload size be, the higher saturation throughput be for error-prone channel. It is identified that there are optimum number of stations corresponding to maximum saturation throughput. The DCF saturation throughput of 802.11a is the highest for error-prone channel. Because system needs 25 dB minimum signal to noise ratio at the data rate of 54 Mbps, this paper used 23 dB and 28 dB as the signal to noise ratio.

We evaluated DCF throughput performance of IEEE 802.11n wireless LAN based on MIMO OFDM with the system parameter defined in Table 12. MCS (Modulation and Coding Scheme) index 15 is used to generate the physical data rate of 130 Mbps with 20 MHz bandwidth and long guard interval. And the two common packets passed down to the MAC layer are 576 bytes (typical size for web browsing) and 1,500 bytes (the maximum size for Ethernet) in length. S(K, ρ, $n_s$, P, n, τ) shows DCF throughput performance over the Rician fading channel. Fig. 20(a) shows DCF throughput on the condition that the subframe' payload size is 576 bytes, the number of stations is 10 and the packet transmission probability is 0.05. In that Fig. 20(a), it is identified most of the the ratio of received average energy per bit-to-noise power spectral density that the larger the Rician factor and the number of subframe are, the better the DCF throughput performance is. Fig. 20(b) has the same condition as Fig 20(a) except the packet transmission probability 0.05 replaced by 0.2. It is identified that if the packet transmission probability is lower, the DCF throughput performance is improved because packet collision probability is decreased. Fig 20(c) has the same condition as Fig. 20(a) except the number of stations 10 replaced by 30. Fig 20(d) is compared to Fig 20(a) about the DCF throughput performance for the subframe size, 576 bytes and 1,500 bytes. It is identified that the larger the subframe' payload size is, the better the DCF throughput performance is.

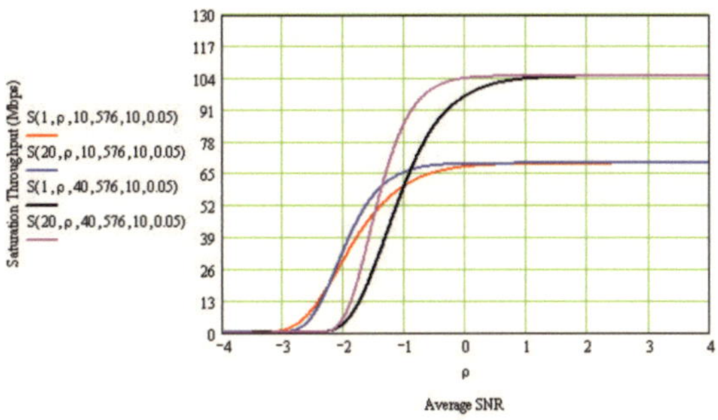


(a) P=576 bytes, n=10, τ=0.05

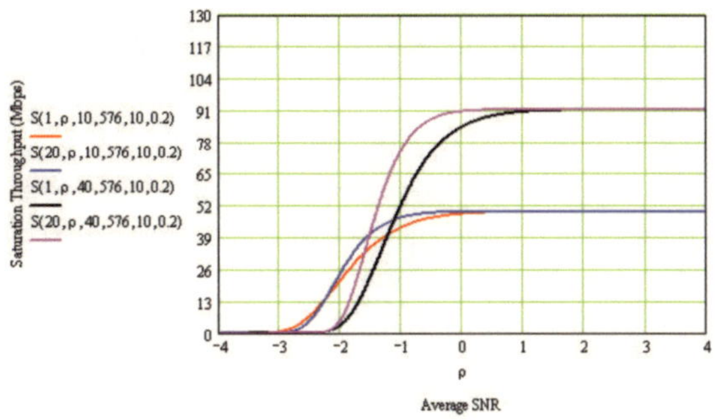


(b) P=576 bytes, n=10, τ=0.2

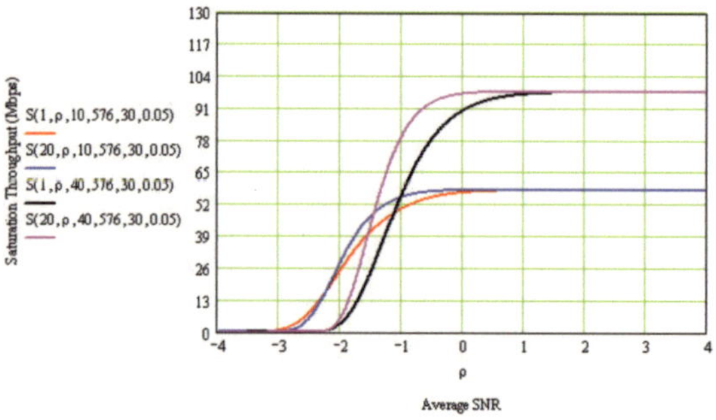


(c) P=576 bytes, n=30, τ=0.05

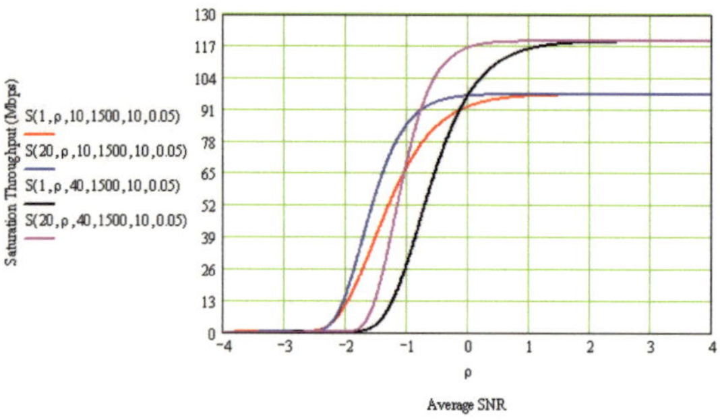


(d) P=1,500 bytes, n=10, τ=0.05

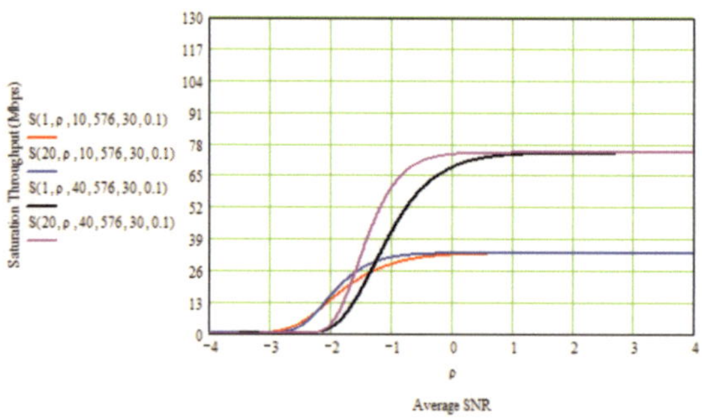


(e) P=576 bytes, n=30, τ=0.1

Fig. 20. DCF throughput in 802.11n (130 Mbps)

#### 5.2.3.2 Analysis of IEEE 802.11a/g/n DCF saturation throughput under flat Rayleigh fading channel with mobile station

This section evaluated DCF throughput of the IEEE 802.11a/g-based mobile LAN with the maximum physical transmission rate of 54 Mbps and that of the IEEE 802.11n-based mobile LAN with the physical transmission rate of 130 Mbps considering 20 MHz MCS (modulation and coding scheme) parameters for two spatial streams, as shown in Fig. 21. Out of the four different physical layers defined in the IEEE 802.11g standard, both ERP-OFDM and DSSS-OFDM standard are only used owing to their maximum transmission rate of 54 Mbps in this evaluation. And the three common packets passed down to the MAC layer are 60 bytes (TCP ACK), 576 bytes (typical size for web browsing) and 1,500 bytes (the maximum size for

Ethernet) in length. In the IEEE 802.11n-based mobile LAN, the number of packets aggregated in one MAC frame varies from 1 to 100, which leads to an aggregated frame's payload length ($L_p$) from 60, 576 and 1,500 bytes to 6, 57.6 and 150 Kbytes. In the Fig. 21(a) ~ Fig. 21(c), the symbol S $(P, \rho, \nu, n, \tau)$ shows the saturation throughput over error-prone channel according to the number of stations($n$) for common packet sizes ($P$) on the condition that packet transmission probability ($\tau$), mobile velocity ($\nu$) and fading margin ($\rho$) are fixed. In the Fig. 21(d) and Fig. 21(e), the symbol S $(ns, P, \rho, \nu, n, \tau)$ and S $(P, ns, \rho, \nu, n, \tau)$ respectively shows the saturation throughput over error-prone channel according to the number of stations ($n$) and the typical number of packets aggregated in one MAC frame ($ns$) for two subframe length on the condition that packet transmission probability ($\tau$), mobile velocity ($\nu$) and fading margin($\rho$) are fixed. For example, in the Fig. 21(a), if the number of stations is 7, packet transmission probability is 0.05, packet length is 1,500 and fading margin is 0.01, mobile station with the speed of 1.25 m/s can get the throughput of 27.238 Mbps, whereas mobile station with the speed of 25 m/s can get the throughput of 26.968 Mbps. In the Fig. 21(d), if subframe length is 30 and the same conditions mentioned above are applied, mobile station with the speed of 1.25 m/s can get the throughput of 113.511 Mbps with six stations, whereas mobile station with the speed of 25 m/s can get the throughput of 84.607 Mbps. Also, Fig. 21(a ~ d) shows that the longer frame (or subframe) length is, the higher throughput is. And, for the same frame (or subframe) length, the higher speed is, the lower throughput is. As the results of evaluation, we also know that there is optimum number of stations to maximize saturation throughput under the error-prone channel. Specially, in Fig 21(e), the number of subframes is considered and it is identified that there is optimum number of subframes to maximize saturation throughput under the error-prone channel.

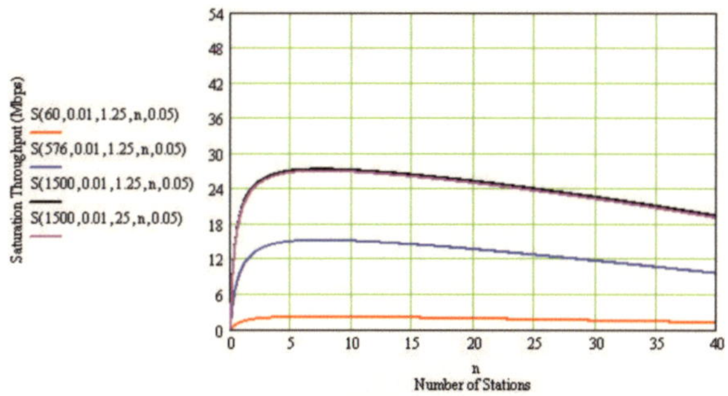


(a) 802.11a OFDM (54 Mbps)

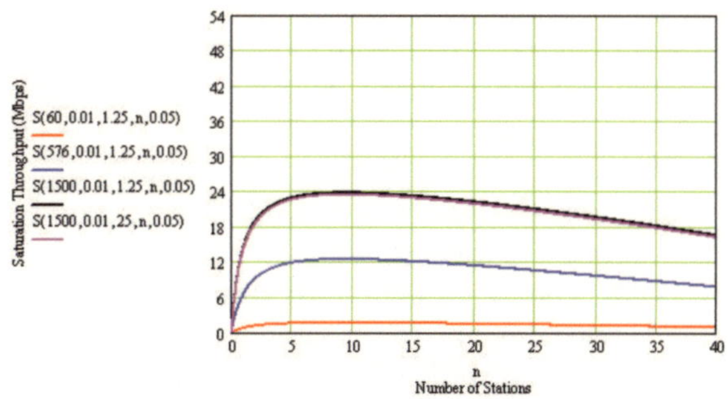


(b) 802.11g ERP-OFDM (54 Mbps)

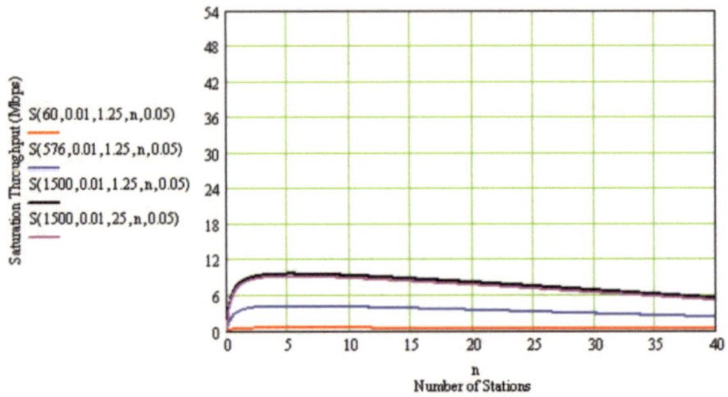


(c) 802.11g DSSS-OFDM (54 Mbps)

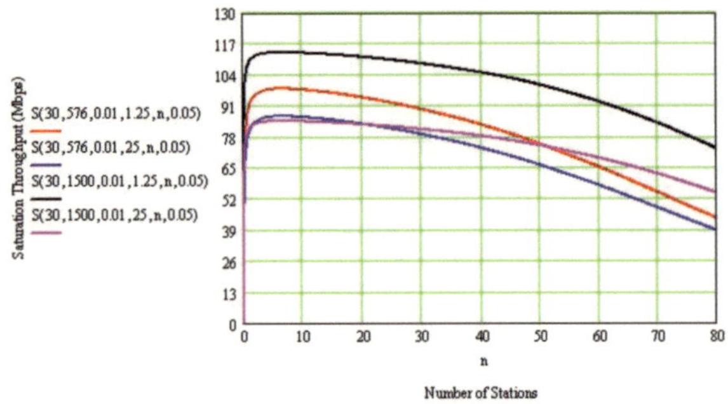


(d) 802.11n OFDM (130 Mbps, number of stations)

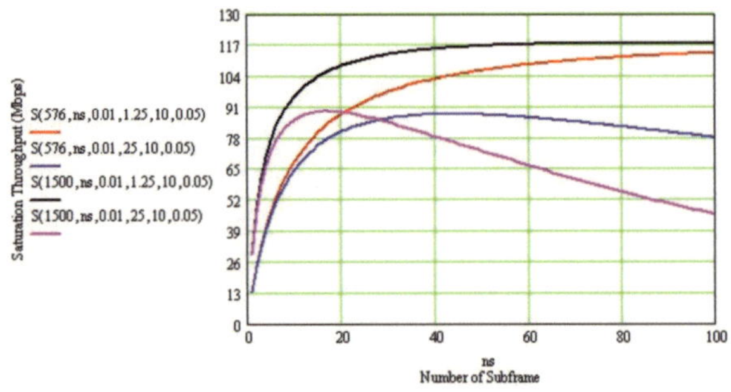


(e) 802.11n OFDM (130 Mbps, number of subframe)

Fig. 21. DCF throughput of IEEE 802.11a/g/n mobile LAN[12, 22]

In conclusion, we obtained the fact that there exist an optimal number of stations (or subframes) to maximize the saturation throughput under the error-prone channel. Also, we can identify that the larger payload (or subpayload) size be, the higher saturation throughput be. And if a mobile velocity of station is increased, the throughput is decreased a little. Out of the three different physical layers defined in this analysis with the maximum transmission rate of 54 Mbps, which are 802.11g ERP-OFDM, 802.11g DSSS-OFDM and 802.11a OFDM, The DCF saturation throughput of 802.11a OFDM is the highest at all the channel environments. In the case of 802.11n, because A-MSDU scheme is applied, it is identified that MAC efficiency of IEEE 802.11n is the best out of all four schemes.

## 6. Remarks

This chapter explored the saturation throughput performance of DCF protocol in the IEEE 802.11a/g/n-based fixed and mobile LAN under the error-prone channel. IEEE 802.11a and IEEE 802.11g have the same maximum transmission rate of 54 Mbps, but the DCF saturation throughput of 802.11a is higher than that of 802.11g. Of the two 802.11g standards, DCF saturation throughput of 802.11g ERP-OFDM is higher than that of 802.11g DSSS-OFDM. We are recognizing that a 802.11n-based device can operate with a 802.11 legacy devices, but 802.11a-based device does not operate with a 802.11b/g-based device. So either constructing 802.11a/n-based mobile LAN or constructing 802.11g/n-based mobile LAN have to be considered for interoperability.

## 7. References


[1] Upkar Varshney, "The Status and Future of 802.11-based Wireless LANs," IEEE Computer, Jun. 2003, pp. 102-105.

[2] Guido R. Hiertz, Dee Denteneer, Lothar Stibor, Yunpeng Zang, Xavier Pérez Costa and Bernhard Walke, "The IEEE 802.11 Universe," IEEE Communications Magazine, January 2010, pp. 62-70

[3] Sixto Ortiz Jr., "IEEE 802.11n: The Road Ahead," IEEE Computer, July 2009, pp. 13-15

[4] Eldad Perahia, "IEEE 802.11n Development: History, Process, and Technology," IEEE Communications Magazine, Jul. 2008, pp.48-55

[5] Zuoyin Tang, Zongkai Yang, Jianhua He, and Yanwei Liu, "Impact of Bit Errors on the Performance of DCF for Wireless LAN," *IEEE*, 2002, pp. 529-533.

[6] Yang Xiao and Jon Rosdahl, "Throughput and Delay Limits of IEEE 802.11," *IEEE Communications Letters*, Vol. 6, No. 8, August 2002, pp. 355 - 357

[7] Jangeun Jun, Pushkin Peddabachagari and Mihail Sichitiu, "Theoretical Maximum Throughput of IEEE 802.11 and its Applications," *Proceedings of the Second IEEE International Symposium on Network Computing and Applications (NCA'03)*, 2003.

[8] Dimitris Vassis, George Kormentzas, Angelos Rouskas, and llias Maglogiannis, "The IEEE 802.11g Standard for High data rate WLANs," *IEEE Network*, May/Jun. 2005, pp. 21-26.

[9] Y. Lin and V. W. S. Wong, "Frame Aggregation and Optimal Frame Size Adaptation for IEEE 802.11n WLANs," in *Proc. IEEE GLOBECOM*, San Francisco, CA, Nov. 2006

[10] D.Skordoulis, Q.Ni, H.Chen,A.P.Stephens, C.Liu, and A.Jamalipour, "IEEE 802.11n MAC Frame Aggregation Mechanisms for Next-Generation High-Throughput WLANs," *IEEE Wireless Communications*, vol.15, Feb. 2008, pp. 40-47.

[11] Xi Yong, Wei Ji Bo, and Zhuang Zhao Wen, "Throughput Analysis of IEEE 802.11 DCF over Correlated Fading Channel in MANET," IEEE, 2005, pp. 694-697.

[12] Ha Cheol Lee, "A MAC Layer Throughput over Error-Free and Error-Prone Channel in The 802.11a/g-based Mobile LAN," MICC 2009, Dec. 2009

[13] Jeff Smith, Jake Woodhams and Robert Marg, "Controller-Based Wireless LAN Fundamentals," Cisco Press, 2011, pp. 71-72

[14] Eldad Perahia and Robert Stacey, "Next Generation Wireless LANs," Cambridge University Press, 2008, pp. 130

[15] Thomas Paul and Tokunbo Ogunfunmi, "Wireless LAN Comes of Age: Understanding the IEEE 802.11n Amendment," IEEE Circuits and Systems Magazine, First Quarter 2008, pp. 28-54
[16] Wireless Gigabit Alliance, "WiGig White Paper: Defining the Future of Multi-Gigabit Wireless Communications, July 2010, pp. 2-5
[17] Ryuta Imashioya, Wahyul Amien Syafei, Yuhei Nagao, Masayuki Kurosaki, Baiko SAI and Hiroshi Ochi, " RTL Design of 1.2Gbps MIMO WLAN System and Its Business Aspect," ISCIT 2009, pp. 296-301
[18] Chang Soon Choi, Eckhard Grass and Maxim Piz, "Performance Evaluation of Gbps OFDM PHY Layers for 60-GHz Wireless LAN Applications," IEEE Conference, 2009.
[19] Chi-han Kao, "Performance of the IEEE 802.11a Wireless LAN Standard over Frequency-selective, Slow, Ricean Fading Channels," Master's Thesis, Sep. 2002.
[20] Giuseppe Bianchi, "Performance Analysis of the IEEE 802.11 Distributed Coordination Function," *IEEE Journal on Selected Areas in Communications*, Vol. 18, No.3, pp. 535-547, Mar. 2000.
[21] IEEE Std 802.11n 2009 " Part11: Wireless LAN Medium Access Control(MAC) and Physical Layer (PHY) specifications: Enhancements for Higher Throughput," Oct. 2009.
[22] Ha Cheol Lee, "A MAC Throughput over Rayleigh Fading Channel in The 802.11a/g/n-based Mobile LAN," MESH 2011, Aug. 2011

# 3

# MAC-Layer QoS Evaluation Metrics for IEEE 802.11e-EDCF Protocol over Nodes' Mobility Constraints

Khaled Dridi[1], Boubaker Daachi[1] and Karim Djouani[1,2]
[1]*Laboratory of Images Signals and Intelligent Systems, Paris-East University*
[2]*F'SATI Institute of Technology/TUT University*
[1]*France*
[2]*South Africa*

## 1. Introduction

Although wireless networks suffer from limit bandwidth, higher bit-error rates (BER's), significant amount of delay, and lower security than wired networks, still they have been emerged as an existing technology for the broadband wireless access, like IEEE 802.11 WLAN. Being fair for sharing medium resources considers the main reason for these weaknesses. Furthermore, Quality of Service (QoS) mechanism, in the recent version of the standard, should improve service differentiation among various types of traffic. It challenged to manage collisions and to support channel variation. Beside, wireless networks are more likely to have higher-ranking on flexibility by allowing easy setting up. If the IEEE 802.11 standard family provides the guarantee for connectivity, sufficient local coverage, required security and enough compatibility with the existing technologies, it is highly expected to carry on real-time applications requirements (Andreadis, 2006). Particularly, the EDCF MAC protocol, which improved a set of parameters, defines the classes of priority for the channel access mechanism during the contention-based period (CP). This can subsequently be declined to a variation of network dynamicity. In fact, when a mobile node crosses the overlay area with other connected nodes, the data transfer can be affected during the handoff intervals. The MAC process fails synchronization and it will be considerably corrupted by generating an amount of packets loss. This effect can be highly intensive depending on the increase of node's mobility rate. Consequently, EDCF protocol loses capacity for QoS delivery and can be reverts to a DCF behaviour reached the threshold limit of stability. Reliability analysis of different traffic classes (video, voice and data), without considering both network topology and node's mobility constraint, is not well appropriated. Dealing with this recent constraint, we propose a study which allows to know how EDCF react facing nodes mobility referring to the MAC protocol stability region. The functional analysis allowed us to follow the mobility of the node and identify the high-rated packets loss areas. To reduce this impact, we specify an algorithm, which operates in different network topology, called multi-coverage algorithm for approving the medium access mechanism. This approach can support overlapping adjacent coverage wireless ranges. For performance evaluation, we studied the most common measured metrics: effective throughput, end-to-end delay and jitter bound. To complete the study, we proposed a mobile scenario, between two adjacent and overlapping wireless stations, within three ranges of mobility rates: low, medium and high. Furthermore, we project to present some issues to improve the behaviour of the protocol by correcting session time BS's hand-off association. Evaluation of the simulation results within three modes of mobility combining the main MAC metrics are detailed and summarised as a user's guide based-on traffic priority scheme.

## 2. IEEE 802.11 MAC legacy

The standard WLAN IEEE 802.11 used Best-effort service model built on FIFO queuing mechanism. The access mode is based upon two different access methods; the mandatory Distributed Coordination Function (DCF) operates in Contention Period (CP) and Point Coordination Function (PCF) for the polling during Contention Free Period (CFP) (IEEE Std. 802.11-1999).

### 2.1 DCF

Carrier Sense Multiple Access with Collision Avoidance (CSMA/CA) protocol is used to regulate the access in the shared medium. So, all wireless nodes have simultaneous access to the same channel resources. If a node wants to transmit, it first senses the medium. The frame is transmitted when the medium is idle for at least a DCF inter-frame space (DIFS) period of time. If the wireless medium is busy, the node chooses a backoff time slot, *B*, consisting of a random number within the Contention Window (CW) interval values (0 to *CW*). This counter, according to each station, is decremented by one when the medium is detected idle for at least one DIFS. Now, when the medium is busy, the *B* timer is frozen (the backoff value is paused to the current value till the state of medium will change). It will be reactivated when the medium becomes free for the next DIFS space. The MAC layer frame is transmitted only when the backoff timer reaches the zero bound.

If a node does not receive the acknowledgement (ACK) frame, it is considered that collision has occurred and the contention window, *W*, is doubled, as:

$$W_n = 2^{c+n-1} - 1 \tag{1}$$

Where $n$ is the number of transmission attempts along with the current one for the frame.

$c$ is a constant, which defines the minimum contention window, as:

$$c = log_2(W_{min} + 1) \tag{2}$$

To start a new backoff process, a new backoff time is chosen. Before sending a new frame after a successful transmission, the backoff mechanism is once more activated. When a transmission is successful, the contention window will reset to $W_{min}$.

### 2.2 PCF

This coordination function is related to an Access Point (AP) based network topology. The AP performs as Point Coordinator (PC). So, PCF corresponds to a centralized and polling-based access mechanism. The condition for the coexistence of both DCF and PCF is the support of PCF. Within a super-frame, to start a CFP, the base station transmits a beacon frame. Once CFP is started, PC maintains a list of the nodes which have demanded to be polled for transmitting data and then it sends poll frames to the nodes. On response, the nodes transmit data packets. A Shorter IFS space is introduced between the PCF data frames to prevent interfering with the DCF mode.

### 2.3 Quality of service and the IEEE 802.11 standard

Service differentiation is one of the most required strategies to manage and improve peak-time network congestion of diverse class of traffic which combining voice, video and data flows. As, IEEE 802.11 standard initially provided a wireless transmission operation mode for the closed local network area, it shows very poor performance regarding link utilization during the competitive applications access (Visser & El-Zarki, 1995). To overcome this weakness and satisfy the service performance across the network, Quality of Service (QoS) concept is proposed. It refers to the ability of a network for providing desired handling of the traffic requirements which meets the expectations of the end applications (Mellouk, 2009). If a network supports a set of traffic specifications, as: bandwidth, transmission delay, jitter-bound, data path-loss, etc then it is supposed to support QoS delivery. Several new mechanisms for service differentiation have been proposed (Dongxia & al., 2009) (I-Shyan & Jheng-Han, 2008) (Whe-Dar & Der-Jiunn, 2008). The quality of traffics including video streaming is getting better performance when the characteristics of the wireless networks are taken into account. The earlier IEEE 802.11 standard treats packets of all traffic categories at the same priority level. Therefore, delay-dependent traffics suffer from network congestion and bandwidth variations.

#### 2.3.1 DCF QoS limitation

DCF coordination function is based on "Best-effort" service model. All nodes compete with the same priority for access to the channel. Moreover, time-bounded multimedia applications require strict bandwidth, delay, and jitter guarantee. Accordingly, there is no service differentiation mechanism to specify and offer better service for prioritized applications than the rest of the traffic.

#### 2.3.2 PCF QoS limitation

PCF mechanism itself weakens QoS; Firstly, IEEE802.11a creates a delay of 4.9 ms because this access process uses beacons frames to separate the two access modes CP and PCF. Secondly, all types of traffics must pass through AP. This condition causes decrease in bandwidth. Thirdly, Mac Service Data Unit (MSDU) size is affected by the transmission of data in different sizes, which makes QoS uncertain for remaining CFP period.

### 2.4 Toward supporting QoS in IEEE 802.11

As discussed above, the original IEEE 802.11 has not the capacity for frames differentiating priority rather it offers an equal chance to all nodes contending for the channel access at the same time. The access method in MAC layer for the new IEEE 802.11e is called Hybrid Coordination Function (HCF), it combines functionalities of both DCF and PCF (IEEE Std.

802.11-1999). Eventually, in order to enhance the contention-based access mechanisms during a CP period of the IEEE 802.11, Enhanced DCF Coordination Function (EDCF) was proposed as well.

**Fig. 1.** shows the super-frame of HCF (Dridi & al., 2008). The great challenge was to make sure that EDCF should be well-matched with the old DCF since large number of devices complying with the old standard had been deployed.

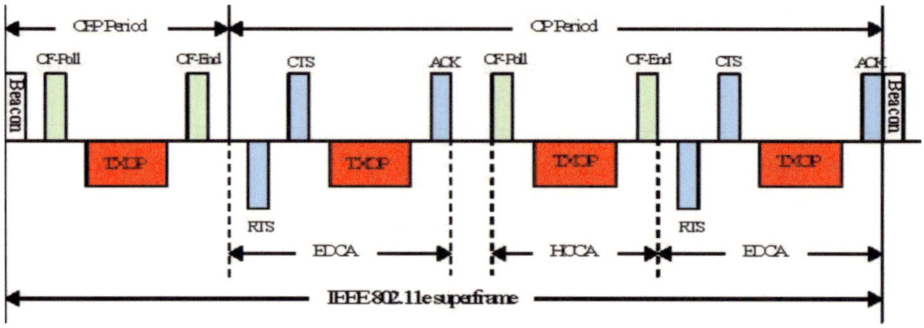


Fig. 1. HCF Super-frame Structure.

The new mechanism classifies the traffic into 8 user classes, with the modified size of contention window ($CW_{min}$) and the inter-frame spaces. Smaller the contention window then shorter will be the backoff intervals. Therefore, the traffic priority will be greater. A new inter-frame space called Arbitration Inter-frame Space (AIFS) is introduced to start decrementing the backoff timer as in ordinary DCF. Besides, AIFS is used to stop waiting a DIFS period of time before trying the access to the medium. AIFS is associated with each traffic class and is evaluated as a DIFS plus a number of time slots. It implies that traffic using a large AIFS will be assigned lower priority. The following scheme in **Fig. 2** depicts the dissimilarities between the IEEE 802.11 coordination functions depending on with vs. without QoS support mechanisms.

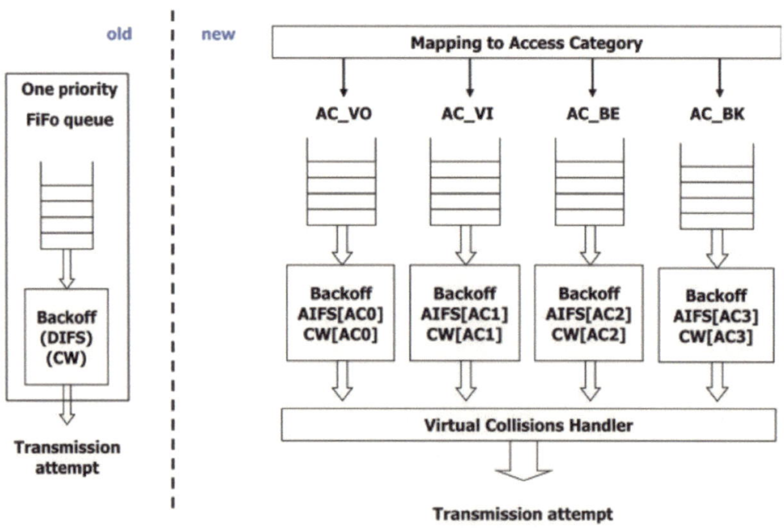


Fig. 2. Advanced QoS improvement mechanism in IEEE802.11e

Toward a better use of the wireless medium, the MAC protocol of IEEE 802.11e should operate in packet bursting mode. It consists of allowing a station to send more frames once it has gained the access to the idle medium through ordinary contention during TXOP-Limit (Dridi & al., 2008). The packet burst is terminated, if a collision occurs or no acknowledgment frame is received, as packet bursting can possibly increase the jitter. The most priority traffic operates with Short Inter-frame Space (SIFS), which is the small time interval between data-frame and Ack-frame.

### 2.5 IEEE 802.11e-EDCF mechanism

To make more efficient for the existing mechanisms of IEEE 802.11, EDCF has been proposed, which aims to enhance the access mechanism by providing the distributed access for the service differentiation. The IEEE 802.11e working group brought an extension to enhance the access mechanisms of earlier standard and provide a distributed access mechanism for service differentiation (Wiethölter & al., 2006). Because a lot of devices have been deployed to improve the DCF, Enhanced DCF (EDCF or EDCA) is the new IEEE appellation. Currently, an intense care aimed to carry on high level of compatibility with the previous generations of the IEEE 802.11 standard.

The MAC protocol of 802.11e standard divides the traffic into eight classes. Each class has different $W_{min}$ and interframe space for the transmission of data. If a node, for example, requires higher priority for data transmission, it would be having smaller $W_{min}$ and hence shorter backoff. If more nodes have the same $W$, the traffic classes are differentiated by having different inter-frame spaces. An inter-frame space called as Arbitration Inter-frame Space (AIFS) is introduced to avoid waiting a DIFS before accessing the medium or like DCF to decrement backoff timer. As mentioned above that EDCF has eight traffic classes with different AIFS but operates with the same DIFS period of time. **Fig. 3.** displays access mechanisms for DCF and EDCF.

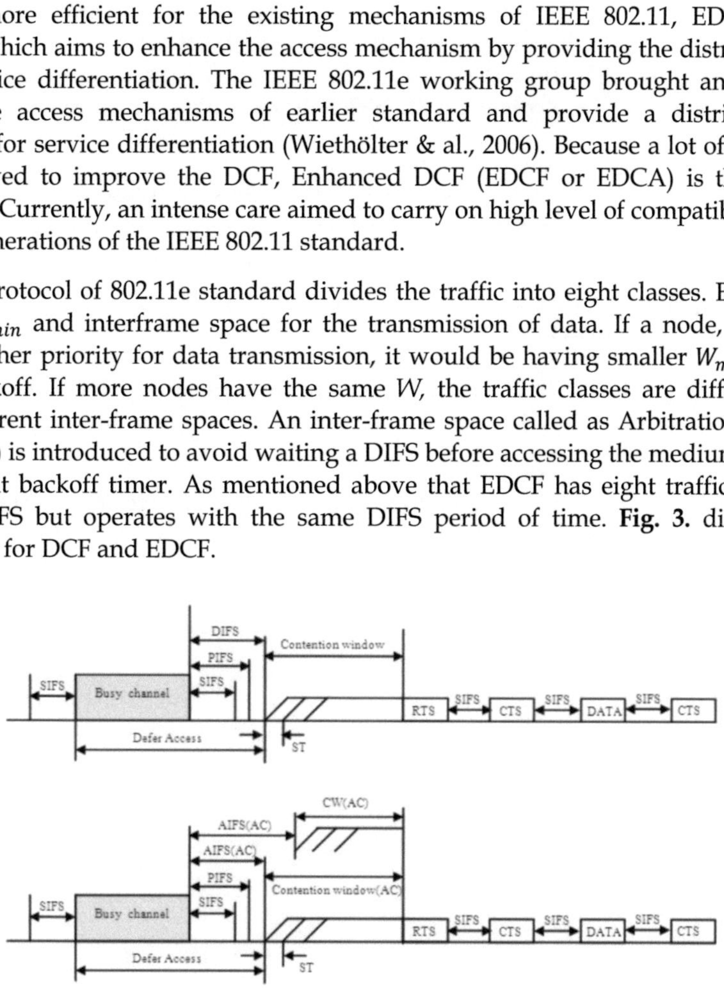


Fig. 3. DCF vs. EDCF access mechanisms.

### 2.6 Mobility

With wireless imbedded devices and high requirement for spreading data transferring, offer facilities for several applications aiming to investigate and control WLAN networks.

Moreover, new connected devices as: PDA, smartphones and tablet are continuously constrained by the mobility. Embedded systems in ground transport (during walk, on cycle, on car, etc) can disturb communication systems which can not follow the small devices. The increase of more requirement for internet keeping a high level of QoS and security, make the wireless network as a big challenge. That's why; we proposed to evaluate the IEEE802.11, as a wide-spreading known network. We focus on the study of the EDCF QoS support mechanism constrained by the node's mobility. We present the analysis of EDCF behaviour through the measurement of the MAC layer metrics: effective throughput, End-to-End delay and jitter.

To avoid several outside varying parameters according to the fact of mobility, we operate on fixed topology scenario while a node in mobility, as a user, connects to the first BS1, passes through the range of BS1, disconnect from BS2 when it will be out of range, detect the coverage of the second BS2, connect with BS2 and keep data transmission (Walke & al., 2006).

Depending on the node's velocity and the EDCF behaviour in QoS stability region, we suggested three ranges of node's mobility domains:

- Low Mobility [5-12] m/s,
- Medium Mobility [15-30] m/s,
- High Mobility [40-60] m/s.

Exceeding the range of 60 m/s, the EDCF will be unsteady and high level of loss-packets is generated. No detection of any kind of QoS delivery which present the limit of the proposal study.

To manage data transmission between wireless stations BSn, an algorithm depending on the position and the quadratic neighborhood distance is proposed. It operates for controlling the data connection with BS's during the node's mobility. The Multi-coverage algorithm is given as follows:

*An example of two BS's network topology*

$m_n(x,y) \leftarrow mobile\ node$
$b_{s_1}(x_1, y_1) \leftarrow base\ station\ 1$
$b_{s_2}(x_2, y_2) \leftarrow base\ station\ 2$
$\check{R}_1 \leftarrow range\ of\ b_{s_1}$
$\check{R}_2 \leftarrow range\ of\ b_{s_2}$
$d_1 \leftarrow |m_n\, b_{s_1}|$ // distance ($m_n$ -- $b_{s_1}$)
$d_2 \leftarrow |m_n\, b_{s_2}|$ // distance ($m_n$ -- $b_{s_2}$)
$d_1 \leftarrow \sqrt{(x - x_1)^2 + (y - y_1)^2}$
$d_2 \leftarrow \sqrt{(x - x_2)^2 + (y - y_2)^2}$
$\boldsymbol{for}\ \{m_n(x,y) \subset (\check{R}_1 \cap \check{R}_2)\}$
$\boldsymbol{if}\ (d_1 <= d_2)$
$then\ connect\ m_n\ to\ b_{s_1}$
$\boldsymbol{else}$
$connect\ m_n\ to\ b_{s_2}$
$\boldsymbol{for}\ \{m_n(x,y) \not\subset (\check{R}_1 \cap \check{R}_2)\}$

$\boldsymbol{if}\ \{m_n(x,y) \subset \check{R}_1)\}$
$then\ connect\ m_n\ to\ b_{s_1}$
$\boldsymbol{else}$
$\boldsymbol{if}\ \{m_n(x,y) \subset \check{R}_2)\}$
$then\ connect\ m_n\ to\ b_{s_2}$
$\boldsymbol{for}\ \{[m_n(x,y) \not\subset \check{R}_1]\ \&\&\ [m_n(x,y) \not\subset \check{R}_2]\}$
$then\ no\ b_s\ connection$

## 3. MAC-layer metrics

A *metric* is a parameter value assigned to each wireless link (path or route) to be used by the respective algorithm or protocol. When the protocol starts processing, this value is measured and evaluated for selecting the better available link to be allocated for data transmission. EDCF protocol is based on shared bandwidth. As we have different composition of traffic, three kind of metric are selected. In each instance of simulation scenario, the values of the metrics are calculated, saved and updated for following continuously the variations of the protocol behaviour. To complete the study, we combine different mobility levels with the main MAC-layer metrics: throughput, end-2-end delay and jitter.

### 3.1 Effective throughput

The amount of data in (bits or Bytes) successfully transferred from source to destination (or it can be measured on hop-to-hop network points), is called throughput. Throughput actually acts as a gauge to measure the amount of data successfully transferred from the source point to the destination point during a specific period of time. The units bits per second (b/s), Bytes per second (B/s), frames per second (f/s), and symbols per second (s/s) are the typical measuring variety of throughput.

$$Throughput = \frac{\sum_n Received\ Packets}{\sum_t Time}\ (B/s) \tag{3}$$

To estimate this metric, only the received packets values are considered in our computation.

### 3.2 End-2-end delay (*E2ED*)

*E2ED* is the time that a packet spends from source to destination. E2ED or latency can rightly be considered as propagation delay between two network points without any extra time processing involved, as the packetization delay at the generator or the packet's analyzer at the destination.

$$E2ED\ = Arrival\ Time - Departure\ Time\ (s) \tag{4}$$

Another possibility can be used to evaluate the E2ED delay; it is obtainable by the average value of a round-trip-time calculation.

$$E2ED_{avg}\ = \frac{\sum_n RTT_n}{n}\ (s) \tag{5}$$

### 3.3 Jitter formulas

Variation in the end-2-end delay measured, either at source or destination, for the corresponding packets is called jitter. The time-sensitive applications, as video or voice streaming, are affected due to the high variations in the jitter. This is why jitter can better measure the performance of traffic-sensitive applications.

Depending upon the requirements, there are numerous ways to measure the jitter.

#### 3.3.1 Jitter combined with E2ED

Here, jitter is measured by calculating the difference in the maximum and minimum E2ED. Most of the equipments deployed up-to-date provide these two delay values. Accordingly, it considered the easiest and the most convenient method for evaluating jitter.

$$Jitter = Max\ E2ED - Min\ E2ED\ (s) \tag{6}$$

#### 3.3.2 Jitter relating to arrival time of packets

Here, jitter is measured by evaluating the difference in the arrivals time of two packets in the destination point instead of *E2ED* values.

$$\delta a_n = a_n - a_{n-1}\ (s) \tag{7}$$

Where, $n$ shows the current packet, $a_n$ is the arrival time of the current packet and $a_{n-1}$ *is* the arrival time of the previous packet.

$$Jitter_n = \delta a_n - \delta a_{n-1}\ (s) \tag{8}$$

#### 3.3.3 Jitter relating to successive packets time

Here, jitter is calculated by measuring the difference in the E2ED of current packet and the previous packet.

$$Jitter_n = E2ED_n - E2ED_{n-1}\ (s) \tag{9}$$

E2ED can also be expressed as:

$$E2ED_n = a_n - d_n\ (s) \tag{10}$$

As: $a_n > d_n$, $E2ED_n$ is always positive,.

From **Eq.(4)**, jitter can be rewritten as:

$$Jitter_n = (a_n - d_n) - (a_{n-1} - d_{n-1})\ (s) \tag{11}$$

$$Jitter_n = (a_n - a_{n-1}) - (d_n - d_{n-1})\ (s) \tag{12}$$

$$Jitter_n = \delta a_n - \delta d_n\ (s) \tag{13}$$

If we look closely, the jitter measured in the third method with **Eq.(9)** is similar to the **Eq.(6)** of the second method with $\delta d_n$ as an additional factor. This later is obviously related to the difference between the departure times. If the source point sends the packets with varying rate, the term $\delta d_n$will compensate this error.

### 3.3.4 Jitter relating to current and average packets delays'

Here, jitter is measured by computing the difference in the E2ED of n<sup>th</sup> packet. It is worth noting that average *E2ED* is measured over the required duration of analysis.

$$Jitter_n = E2ED_n - E2ED_{av}\,(s) \tag{14}$$

## 4. Scenario and network simulations

### 4.1 Setting mobility on wireless LAN

We started to analyze the impact of the network variations topology and dynamicity on the stability of EDCF mechanism using NS-2 a discrete event simulator targeted at networking research (Fall & Varadhan, 2011). Results of the comparison study between static scenarios, took as a reference and dynamic scenario show the degree of sensitivity of QoS service delivery to the mobile environment. The degradation rate is important where the EDCF can reach the instability region. Consequently, it loses significantly its service differentiation quality and looks like DCF in the worst case of simulation, as shown in **Fig. 4** and **Fig. 5.** The present study focuses on the stability region of EDCF under mobility constraint. QoS scheme is evaluated depending on three domains of velocity (Low, Medium, and High) through the set of MAC-layer metrics.

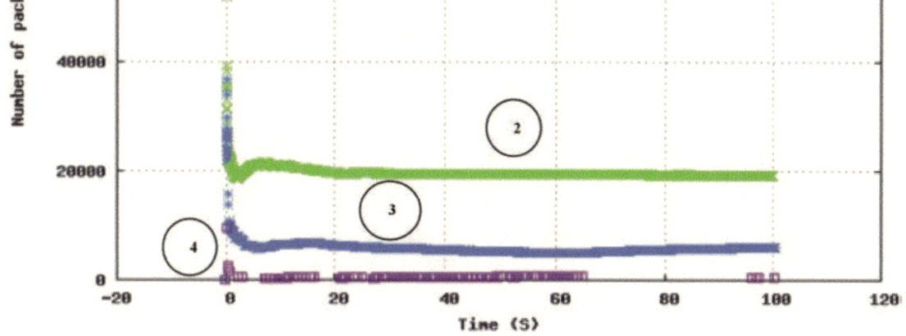


Fig. 4. Throughput without mobility

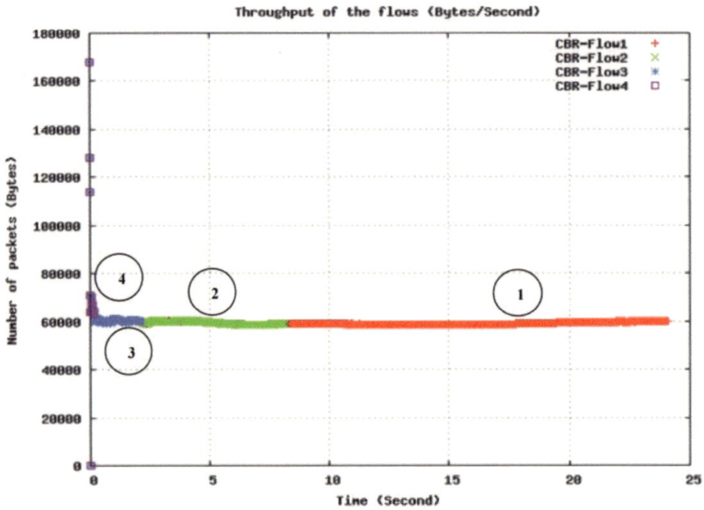


Fig. 5. Throughput with mobility

We proposed a scenario within hybrid network, which is composed of wireless and wired nodes, which can communicate through Base Stations (BS's, $b_{s_1}$& $b_{s_2}$). We specified a solution with two BS's observe that what happens when a Mobile Node ($m_n$) moves out from one BS to another (**Fig. 6**). At this time, EDCF mobility behaviour can be well evaluated.

For the different Access Categories (AC's) of service and to avoid TCP control packets exchanges', 4 CBR-traffics based on UDP transmission protocol are used.

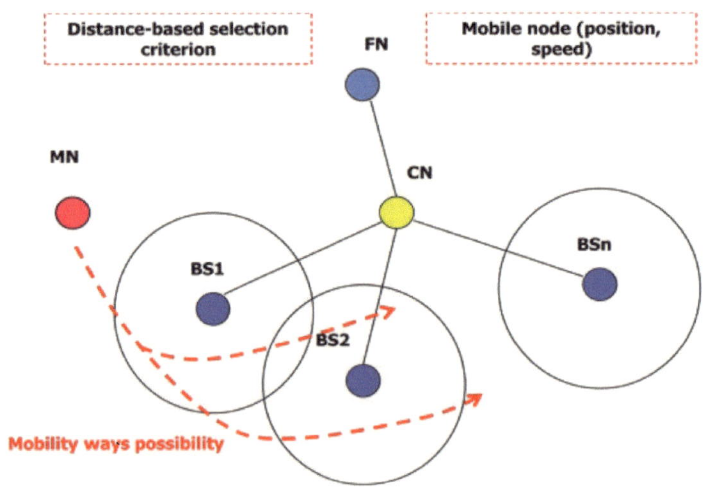


Fig. 6. Mobility scenario through multiple BS's handoff

### 4.2 Environment and simulation parameters

Within a table, the environement of the proposal scenario and the parameters used for developing the mobility scheme are presented (**Tab. 1**).

| | |
|---|---|
| Physical radio-propagation model | Two Ray Ground |
| MAC-level type on Data-Link layer | IEEE 802.11e |
| Queuing class interface | Priority Queuing strategy |
| Max queue length | 50 packets |
| Routing Protocol | DSDV |
| QoS Access Categories | AC0, AC1, AC2, AC3 |
| Wireless Data-Rate | 1Mb (80% shared between AC0 & AC1) |
| Mobile node & Network topology | MN, BS1, BS2, CN, FN |
| Wired to Base-Station links | 5Mb / 2ms / FIFO-DropTail queuing scheme |
| Development tools: | Win-32 on Cygwin environment<br>NS-2, Nam, trace-file format, Xgraph<br>ConTXT program editor, Gnuplot |

Table 1. NS-2 simulation parameters & development tools

## 5. Results analysis

To focus only on the impact of mobility, the present study doesn't take care of the fading effect related to the wireless channel and the non-stationary flows in high data-rate. 5 Mbps are supported by the wired links, and 1 Mbps of data-rate in wireless medium is shared by the different classes of traffic.

As our work focuses on MAC layer, so, without making any comparison of network layer protocols, we used the same Destination Sequenced Distance Vector (DSDV) routing protocol for all the simulation scenarios. To appear congestion event, we preferred limiting the size of the queues at 50 packets. This can allow dropped packets generation, as assuming in a real network, specially when the scheduler is out of its own capacity (Dridi & al., 2010).

### 5.1 First case: EDCF – Behaviour over "low mobility" domain

#### 5.1.1 Throughput

During low mobility and connection with the BS1, throughput achieves levels 60 Kbps and 20 Kbps for AC0 and AC1 respectively. These tow traffics share 80% of total bandwidth is depicted in (**Fig. 7**).

During connection with the BS2, the traffic switches to the half level of the top start throughputs (30Kbps & 10Kbps) and increases during the range of mobility. In this class of mobility, there is no rapid saturation event and the EDCF has enough gap to follow the mobility rhythm by increasing the level of the throughput application adequately. By no need of high throughput network resource, VOIP applications can well transmitted without be thresholded by saturation. QoS is maintained and the behaviour remain the same in both ACs traffic categories with a slight rapid has expected in AC1 to reach the top (in the border of 12 m/s).

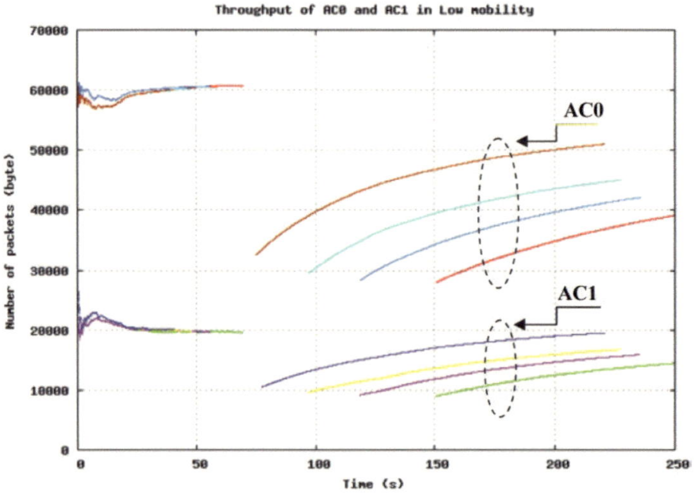


Fig. 7. Throughput for low mobility

### 5.1.2 End-to-end delay

During the connection with BS2, the AC0 stays around 0.2s, this is very convenient for real-time and almost delay-dependent applications, like audio and video streaming. Unfortunately, EDCF shows this capability only for the AC0 class. On the other hand, during connection with BS2 in AC1, EDCF shows some weakness and remains still sensitive to the fact of mobility, as depicted in the graph in (**Fig. 8**), level of 0.8s is reached for several times. The limit of 0.2s for a reasonable VoIP conversation is exceeded. Fortunately, the value of 0.5s is rapidly bounded which can be welcomed for CBR-MPEG video stream applications.

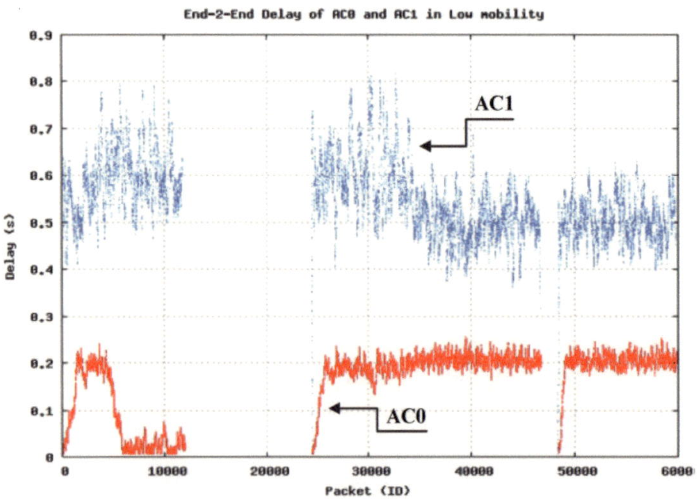


Fig. 8. E2ED for low mobility

### 5.1.3 Jitter

The graph in the **Fig. 9** shows the jitter plotted in two sides of x axis. The negative peaks for packets arrive early, and positive peaks for packets arrive late. The variation of the intensity of jitter in both sides identifies the quality of transmission (ex. degradation quality during a call in VoIP).

For AC0, the jitter is bounded around 0.01s for the all transmissions. An interesting behaviour of EDCF is observed at the negative side. The flow stays quite steady before and after the switching period between the BS's. It shows the ability of the EDCF scheduler to track jitter measurements and adjust buffer size to reduce the jitter impacts. Unfortunately, as we can see in the same graph, this behaviour raises the jitter impact of AC1 (it can reach 0.06s with 0.12s peak-to-peak of variance). Outside this region, EDCF with CBR flow is quite steady.

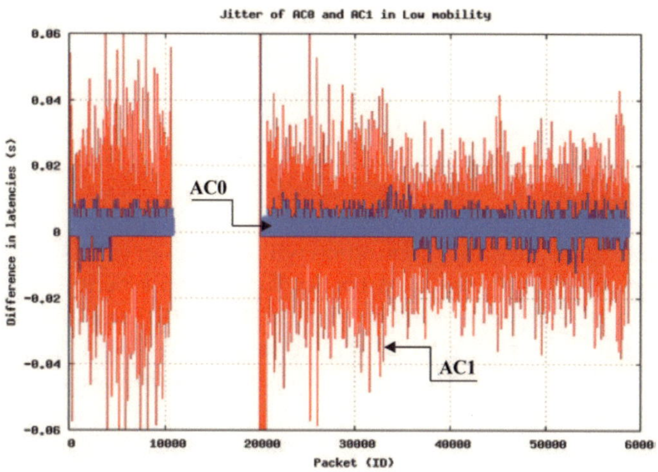


Fig. 9. Jitter for low mobility

## 5.2 Second case: EDCF – Behaviour over "medium mobility" domain

### 5.2.1 Throughput

At the start during connection with BS1, the top AC0 and AC1 throughputs stay unchanged. In AC0 traffics with the increase of the node's mobility, the throughput slightly increases. Comparing to the previous mode, the graph for this case (**Fig. 10**) and previous case are much closer and can't grow over 57 Kbps (start of saturation bound). Even the start top level is not reached; EDCF is not capable to increase throughput for the highest priority traffics. In contrast to AC0, AC1 can gain more flexibility and it increases over the top start level even the curves saturation is expected (more than 20Kbps reached in 30m/s).

In this mobility mode, EDCF guarantees for the maximum throughput. This is highly required variation for the throughput sensitive application (VBR flow) (Rong & Xuming, 2009), as they need robustness over user mobility. Comparing the connections with both of the BS's, we can observe that AC1 does not waste the bandwidth between two connections as AC0 does.

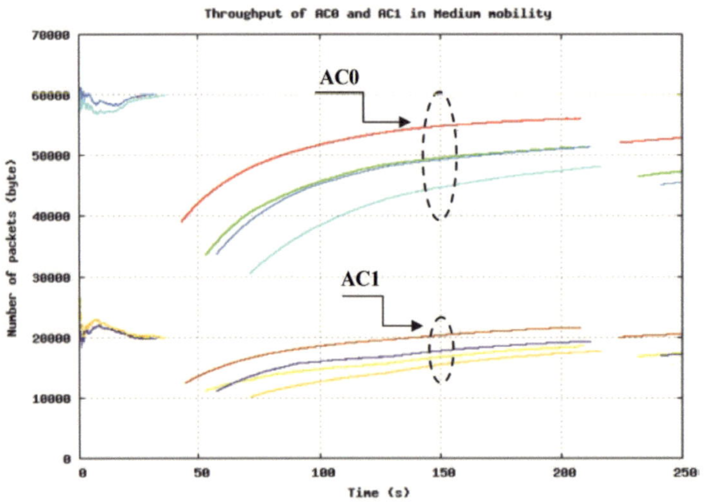


Fig. 10. Throughput for medium mobility

### 5.2.2 End-to-end delay

AC0 delay gains more stability after connection with BS2 down to 0.1s, and up at ~0.2s. This zone can be reserved for high sensitive traffics for the brief period of time (as in Burst mode). The AC0 in the graph depicted in **Fig. 11**, shows that less than 30000 packets are allowed for one burst. AC1, after establishing the connection (0.5s) attains an average delay and stays steady for the rest of transmission. This is the most important characteristic of this mode that it can bring to high priority traffic in the same transmission with a small shift of delay but granting a maximum stability. We find the best result at 30 m/s.

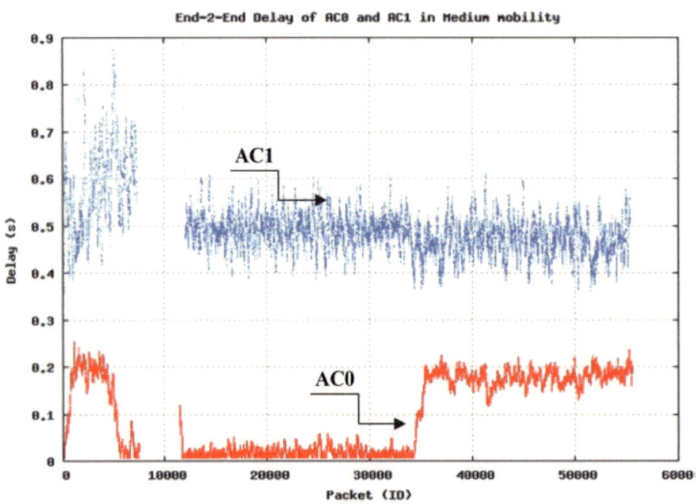


Fig. 11. E2ED for medium mobility

### 5.2.3 Jitter

As shown by **Fig. 12**, high stability of the negative side of AC0 (before the range of 35000 packet ID) shows that there are no packets coming early. AC1 proves E2ED stability with a reasonable level of jitter (<0.5s) after connection with BS2, with 0.1s peak-to-peak of variance. Almost all of the traffics can support the range of mobility of this mode with a condition of burst (as explained previously).

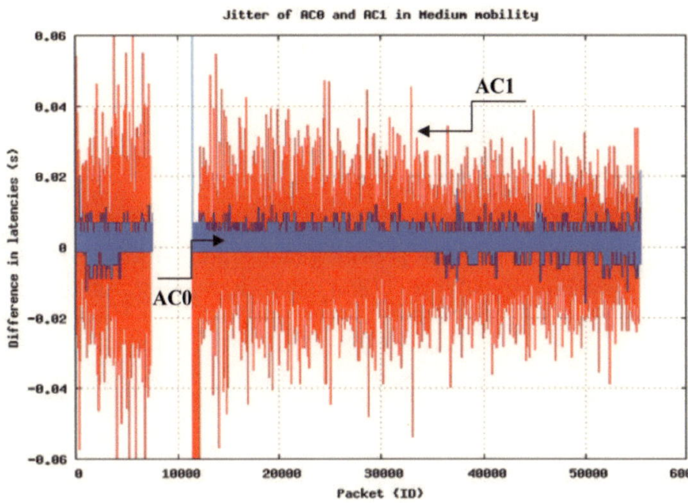


Fig. 12. Jitter for medium mobility

## 5.3 Third case: EDCF – Behaviour over "high mobility" domain

### 5.3.1 Throughput

Within this mode, AC0 and AC1 keep the same behaviour independently to the node's mobility, as shown in **Fig. 13**. EDCF lasts the throughput level comparing to the other modes (15 Kb/s, 30 Kb/s for AC0 and AC1 respectively). It cannot follow the rhythm of the mobility to adjust the throughput accordingly. This behaviour is mainly involved to the size of buffer of the scheduler which is not able to support higher speeds (40 m/s as a critical). All of the traffic classes are penalized with fixed level of throughput, depending on the defined buffer parameters (the size of queue and the used queuing scheduling strategy) (Rong & Xuming, ICCSN-2009).

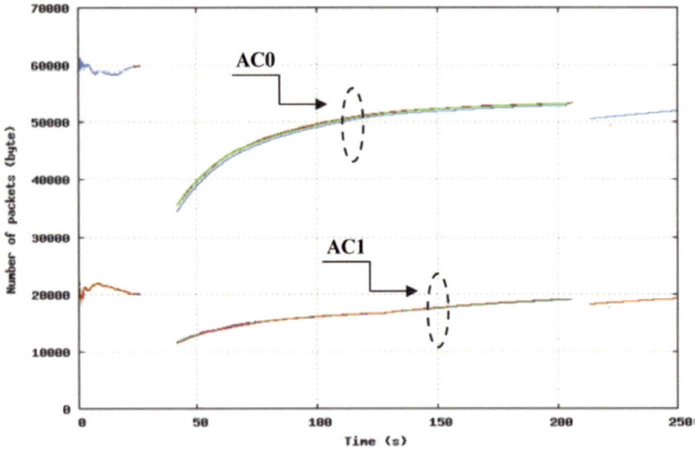


Fig. 13. Throughput for high mobility

### 5.3.2 End-to-end delay

AC0 stays in fixed position (0.2s) as the previous mode. AC1 shifted with high delay (0.9s) and stays in the same level independently of the network mobility. As we can see in (**Fig. 14**)

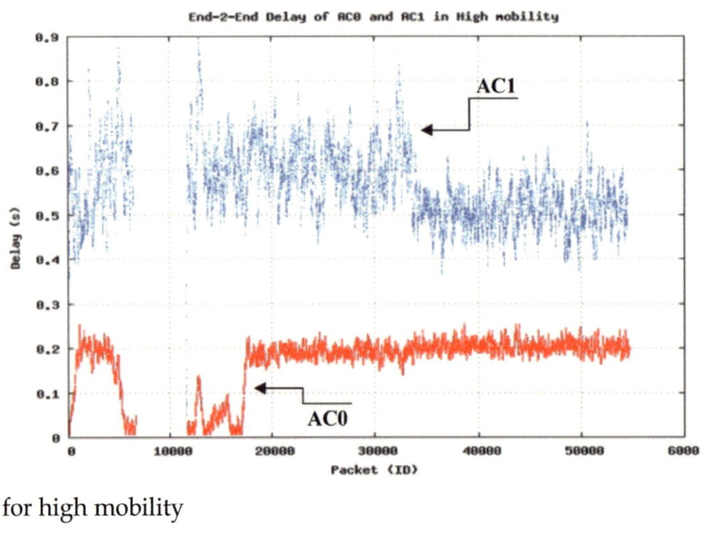


Fig. 14. E2ED for high mobility

### 5.3.3 Jitter

AC0 is in the worst case than the medium mode, as stability is decreased (less than 10000 packets for burst). AC1 keeps the same levels of jitter (0.06s with 0.12 peak-to-peak of variance in **Fig. 15**) with random concentration is slightly appeared discerned in the path of mobility.

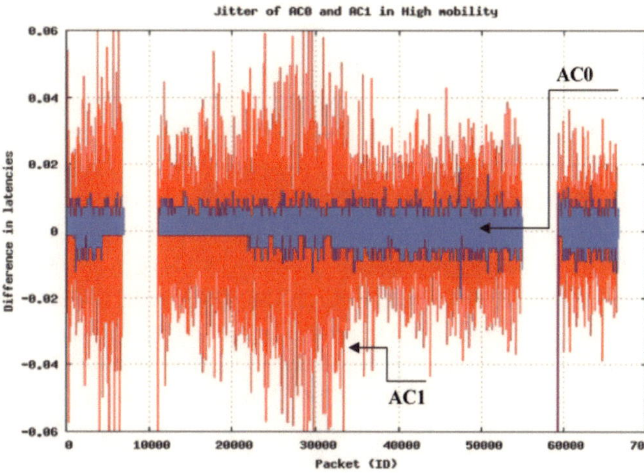


Fig. 15. Jitter for high mobility

The three metrics are worth studying and discussing where the table (**Tab. 2**) Summarizes the behaviour of EDCF under the three mobility domains according to the traffic aware-MAC metric pertinency.

| Mobility/Metric | Low | Medium | High |
|---|---|---|---|
| Throughput | Best ! for real-time applications & video conferencing | Satisfactory for Data transmission | Satisfactory for Emails, SMS, MMS |
| E2ED | Good for all streaming traffics | Satisfactory for CBR-MPEG | Not satisfactory for almost traffic |
| Jitter | Good for VoIP traffic | Good for no-streaming traffic | Worst ! Critical level of dropping packets |

*High performance*

*Poor performance*

Table 2. EDCF behaviour for the MAC metrics constrained by three levels of Mobility

## 6. Conclusion

The proposed study of this chapter focuses on the performance of the EDCF protocol under the node's mobility constrains. The IEEE 802.11 standard reveals high sensitivity to the nodes' position and velocity. These can significantly decrease the standard QoS service ability. By looking for QoS stability region of the MAC protocol, several tests are performed over the main layer-link metrics; throughput, End-2-End delay and jitter in order to quantify the mobility effect. Therefore, different classes of traffic are defined. We ended the study by proposing a benchmark which summarized the impact of these metrics according to three zones of stability. Furthermore, the QoS mechanism behaved differently depending on the rhythm of mobility apply in each scenario using NS2 network simulator. The study of MAC protocol, even the range is limited by PHY layer, allows extension since it can operate easily within cooperation topology. The approved results can help users to identify the borderline of service's steadiness depending on the requirements of the traffic flow. Following this optimistic approach, the study will be expanded for supporting channel fading effect, multipath distortion and BER vs. SNR links' quality within node's cooperative diversity network.

## 7. References


Walke B. H., Mangold, S., & Berlemann, L. (2006). *IEEE 802 Wireless Systems Protocols, Multi-hop Mesh/Relaying, Performance and Spectrum Coexistence,* John Wiley & Sons Ltd, ISBN: 100 470 01439-3, Chichester, England

Andreadis, A. & Giambene, G. (2003). *Protocols For High Efficiency Wireless Networks*, Kluwer Academic Publishers, ISBN: 1-4020-7326-7, Dordrecht

Mellouk, A., (2009). *End-to-End Quality of Service Engineering in Next Generation Heterogeneous Networks*, ISTE ; Hoboken, NJ : Wiley, ISBN : 978-1-84821-061-5, London, England

Dridi, K., Javaid, N., Daachi, B., & Djouani, K. (2009). *IEEE 802.11e-EDCF evaluation through MAC-layer metrics over QoS-aware mobility constraints,* In Proceedings of MoMM International Conference (December 2009). pp.211~217, Kuala Lumpur, Malaysia

Visser, M. A., & El Zarki, M. (1995). *Voice and data transmission over an 802.11 wireless network,* in: *Proceedings of PIMRC'95,* pp. 648–652, Toronto, Canada (1995)

Dongxia, X., Taka, S., & Hai, L. (2009). *An Access Delay Model for IEEE 802.11e EDCA,* IEEE Transaction on Mobile Computing, VOL. 8, NO. 2, Feb, 2009

Dridi, K., Javaid, N., Djouani, K., Daachi, B. (2010). *Performance Study of IEEE802.11e QoS in EDCF-Contention-based Static and Dynamic Scenarios,* IEEE ICECS International Conference, Computing Research Repository Journals, abs-1012-4113, (Dec. 2010)

I-Shyan, H. & Jheng-Han, W. (2008). *Performance assessment of service differentiation in IEEE 802.11e Wireless LANs,* International Journal of Ad Hoc and Ubiquitous Computing, Vol. 3, No.1, pp. 21 - 32, 2008

Whe-Dar & Der-Jiunn D. (2008). *Service Differentiation in IEEE 802.11e HCF Access Method,* Springer Berlin, Heidelberg, 2008

IEEE Std. 802.11-1999, Part 11: *Wireless LAN Medium Access Control (MAC) and Physical Layer (PHY) specifications,* Reference number ISO/IEC 8802-11:1999(E), IEEE Std. 802.11, 1999 edition

Dridi, K., Djouani, K. & Daachi, B. (2008). *Three Scheduling-levels algorithm for IEEE 802.11e QoS efficiency improvement,* International Conference on Advances in Mobile Computing & Multimedia (MoMM), (December 2008), Linz, Austria

Rong, H. & Xuming, F. -2009). *A fair MAC scheme for EDCA based wireless networks,* Testbeds and Research Infrastructures for the Development of Networks & Communities and Workshops, 2009

Rong, H. & Xuming, F. (2009). *A Fair MAC Algorithm with Dynamic Priority for 802.11e WLANs,* ICCSN, 2009

Wiethölter, S., Emmelma, M., Hoene, C. & Wolisz, A. (2006). *TKN EDCA Model for ns-2,* Technical Report TKN-06-003, Technische Universität Berlin, June 2006

Tehuang, L., Wanjiun, L., & Jeng-Farn, L. (2009). *Distributed Contention-Aware Call Admission Control for IEEE 802.11 Multi-Radio Multi-Rate Multi-Channel Wireless Mesh Networks,* ACM/Springer MONET Special Issue on New Advances in Heterogeneous Networking for Quality, Reliability, Security and Robustness, Vol. 14, No. 2, pp. 134, 2009

Fall, K. & Varadhan, K. (2011). *The NS Manual (formerly ns Notes and Documentation),* UC Berkeley, LBL, USC/ISI, and Xerox PARC. Available from http://isi.edu/nsnam/ns/doc/ns_doc.pdf

# 4

# Techniques for Preserving QoS Performance in Contention-Based IEEE 802.11e Networks

Alessandro Andreadis and Riccardo Zambon
*Università degli Studi di Siena, Dipartimento di Ingegneria dell'Informazione, Siena, Italy*

## 1. Introduction

WLANs are widely adopted worldwide for creating hotspots in university campuses and public or private buildings, mainly because they are easy to install and to use.

The new information and communication technologies should allow users to enjoy current as well as future applications and services in an efficient and effective way, in order to achieve a certain level of user perceived satisfaction. Such applications are mainly based on heterogeneous multimedia flows with different characteristics and requirements. Audio, text, video and voice are the basic components of multimedia flows and they have different needs for efficient delivery.

In order to define the level of degradation of a multimedia traffic traversing a network, such as a movie in streaming modality or a Voice over IP (VoIP) conversation, several Quality of Service (QoS) parameters can be taken into account. These parameters are the throughput, the maximum delay between packet transmission and final delivery (i.e., end-to-end delay), the delay jitter experienced by consecutive packets and the percentage of lost packets in the network, due to network congestion or link corruption.

Furthermore, packets correctly received after their maximum delay requirements must be considered as they were lost, because they are not useful for a satisfactory reconstruction of the flow. Hence, new mechanisms are needed for supporting QoS performance of multimedia flows.

Today, the IEEE 802.11 WLAN standard family represents one of the most widely adopted technologies for delivering different traffics through wireless networks. The diverse requirements of heterogeneous traffics contending the wireless channel heavily affect efficient delivery over WLANs.

In the legacy IEEE 802.11 (IEEE, 1999) standard, QoS support was not envisaged, but new multimedia traffics raised up the need of efficient mechanisms for QoS management; the main answer to this issue is represented by IEEE 802.11e specifications (IEEE, 2005).

IEEE 802.11e allows to improve traffic performance over WLANs (Davcevski & Janevski, 2005 ; Choi et al., 2003), by introducing an enhanced MAC scheme. However, further improvements are still possible in this direction and several modifications have been proposed in literature to increase the performance and efficiency of the resource allocation mechanisms envisaged by 802.11e (Thottan & Weigle, 2006 ; Nafaa, 2007).

This chapter deals with the MAC contention phase of IEEE 802.11e; in particular, we aim at providing a comprehensive overview on the state of the art about techniques for enhancing performance and for preserving QoS requirements under heavy traffic conditions. Advantages and disadvantages of each technique will be exhaustively described.

In the second section, we provide an overview of the IEEE 802.11e MAC scheme, in order to understand the basics of this technology, with respect to channel access methodologies.

In the third section, we introduce the main research works related to performance enhancement through the manipulation of standard parameters. Many of these works deal with the problem of fair resource allocation in uplink and downlink directions in infra-structured WLANs.

Under heavy traffic conditions, the fairness problem can be faced through a correct dimensioning of parameters, so as to modify the resource allocation ratio between the Access Point (AP) and the wireless stations.

The efficiency increase which can be achieved by adopting the described techniques remains somewhat limited by WLAN resource availability; therefore, beyond certain traffics loads, an admission control algorithm results necessary for preserving QoS performance of existing flows in terms of throughput, delay, jitter and packet loss, and to avoid that the ingress of new flows seriously damages QoS performance of active traffics.

For this reason, the fourth part of this chapter discusses the main admission control algorithms for WLANs, by comparing and classifying the most recent works in this research area. Categorization of admission control schemes will be carried out within three main classes: measurement-based, model-based, and joint measurement and model-based admission control scheme. The main peculiarities of these schemes will be discussed.

Finally, open issues and conclusions are described in section 5 and 6 respectively.

## 2. The IEEE 802.11e MAC scheme

The original IEEE 802.11 standard handles traffic on a best-effort basis, through the *Distributed Coordination Function* (DCF) MAC scheme. This means that all frames are equally treated and, therefore, DCF is not suitable for managing real-time and interactive multimedia services.

For this reason, an enhancement to DCF has been introduced by 802.11e specifications for QoS deployment. The enhanced MAC scheme, named *Hybrid Coordination Function* (HCF), operates with two access modes (802.11e, 2005):

- *Enhanced Distributed Channel Access* (EDCA), which defines a new contention-based channel access scheme;
- *HCF Controlled Channel Access* (HCCA), which defines a contention-free channel access scheme.

This chapter is limited to describe QoS enhancement which could be achieved through EDCA, since this scheme is mandatory and supported by all 802.11e devices. HCCA is still rarely implemented, since it has a more complex hardware implementation (Banchs et al., 2005).

The basic mechanism of EDCA scheme is based on *Carrier Sense Multiple Access with Collision Avoidance* (CSMA/CA), where collision avoidance is realized through a random back-off time before transmission. In particular, when a station has frames ready for delivery, it starts contending the channel. After the station successfully detects that the wireless channel has been idle for a certain time interval, it chooses a random back-off time within a Contention Window (CW). Frame transmission can start only if the station detects an idle channel for this additional random amount of time.

EDCA introduces differentiated channel access probabilities to frames contending for channel resources. Four Access Categories (AC) are implemented at each station supporting QoS facilities (QSTA), being it an Access Point (QAP) or a wireless host (Figure 1).

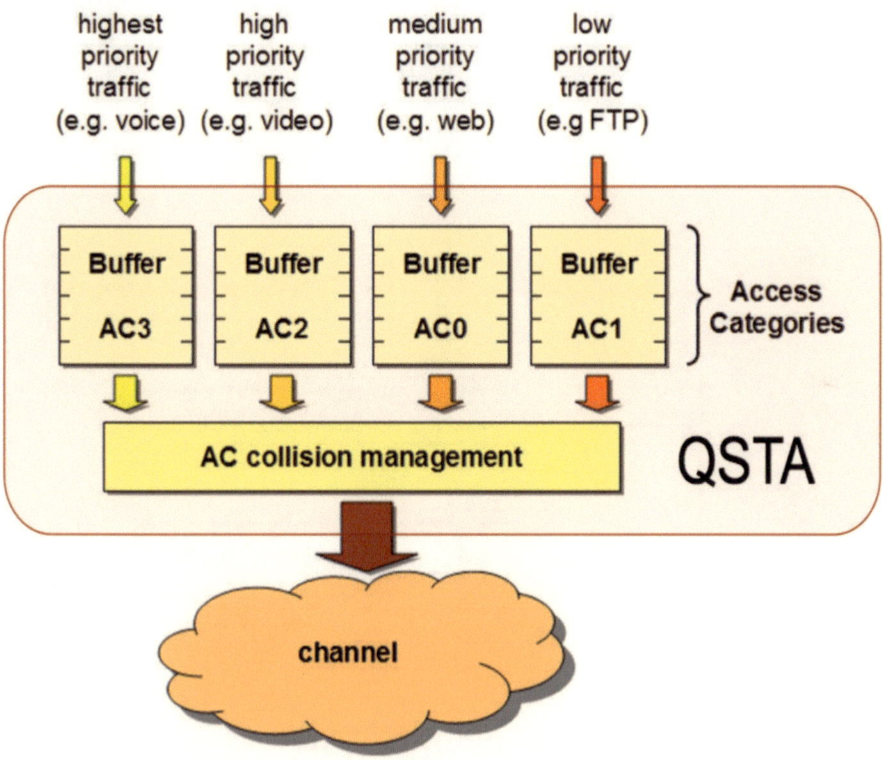


Fig. 1. Internal buffers for frame differentiation inside a QoS enabled station

Prioritization of the different traffics (e.g., voice, video, data) is realized by mapping frames to the proper AC, according to their QoS requirements, and by assigning to each AC an appropriate set of four EDCA parameters (802.11e, 2005). These parameters are used for regulating the channel contention phase and they are periodically broadcasted by the QAP through the beacon frame.

In particular, EDCA parameters are the following:

- *AIFS (Arbitrary Inter-Frame Space)* - this is a waiting time interval, named *AIFS[AC]*, whose duration is different for each AC, so as to take into account its QoS requirements. According to the CSMA/CA scheme, before starting its back-off timer, each station must detect an idle channel for at least *AIFS[AC]*. Actually, the shortest waiting time defined by 802.11e MAC is *Short Inter-Frame Space* (SIFS) and it is adopted for highest priority frames, which are short control messages, such as acknowledgments (ACKs) of data packets. Consequently, the waiting time for data frames belonging to a specific AC, *AIFS[AC]*, must be greater than SIFS and it is calculated as:

$$\text{AIFS[AC]} = \text{SIFS} + \text{AIFSN[AC]} \times \text{SlotTime} \quad (1)$$

  where *AIFSN[AC]* is an integer number greater than zero and *SlotTime* depends on the PHY in use (e.g., 20 μs for 802.11b).
- *CWmin* and *CWmax* (Contention Window parameters) - they regulate the minimum and maximum dimension of the random back-off window size. In particular, the window size is firstly initialised at *CWmin*, but after each collision the maximum back-off window size is doubled up, with an upper bound of *CWmax*. The QSTA must randomly choose its random back-off timer, less than or equal to the current value of CW (Figure 2).

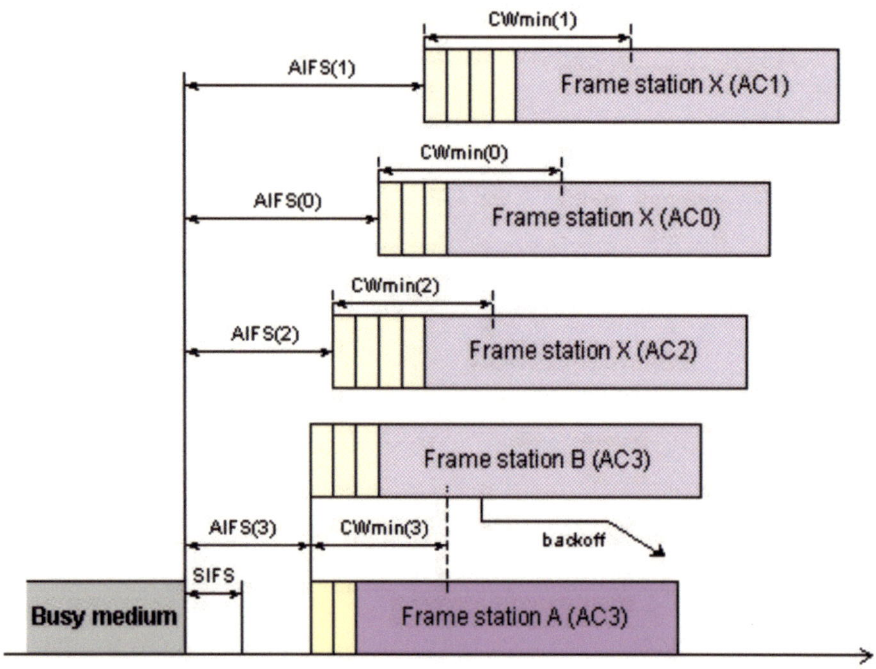


Fig. 2. EDCA contention access scheme

- *TXOPlimit* (*Transmission Opportunity limit*) - when a station wins a channel contention, it can transmit a burst of frames so as to improve MAC efficiency in case of small packets. *TXOPlimit* represents the maximum time interval a station is allowed to use the channel for transmitting a burst of frames, without the need to enter again in channel contention phase. The zero value allows the transmission of a single frame per channel access (Figure 3).

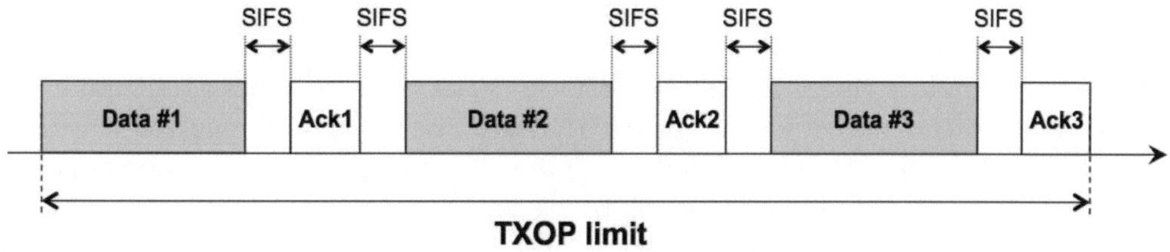


Fig. 3. *TXOPlimit* for the transmission of a burst of frames

Table I reports, for each traffic category, the EDCA parameter values indicated by the standard. VoIP traffic has the highest priority and it is assigned to AC3, video is associated to AC2, best effort (e.g., web packets) corresponds to AC0 and background traffic (e.g., ftp) has the lowest priority and it is mapped into AC1.

| Traffic type | AC | Priority | AIFSN | CWmin | CWmax | TXOPlimit |
|---|---|---|---|---|---|---|
| VoIP | AC_VO (AC3) | 3 | 2 | 7 | 15 | 3.264 ms |
| Video | AC_VI (AC2) | 2 | 2 | 15 | 31 | 6.016 ms |
| Best effort | AC_BE (AC0) | 1 | 3 | 31 | 1023 | 0 |
| Background | AC_BK (AC1) | 0 | 7 | 31 | 1023 | 0 |

Table 1. EDCA parameter set indicated in IEEE 802.11e specifications

At the beginning of each beacon interval (usually 100ms), the QAP broadcasts a beacon frame containing control information for QoS management. In particular, the "*EDCA Parameter Set*" element (Figure 4) of the beacon frame is used to send the EDCA QoS parameters to the wireless stations. Specifically, these parameters are contained inside the "*Parameter Record*" field of each AC (802.11e, 2005).

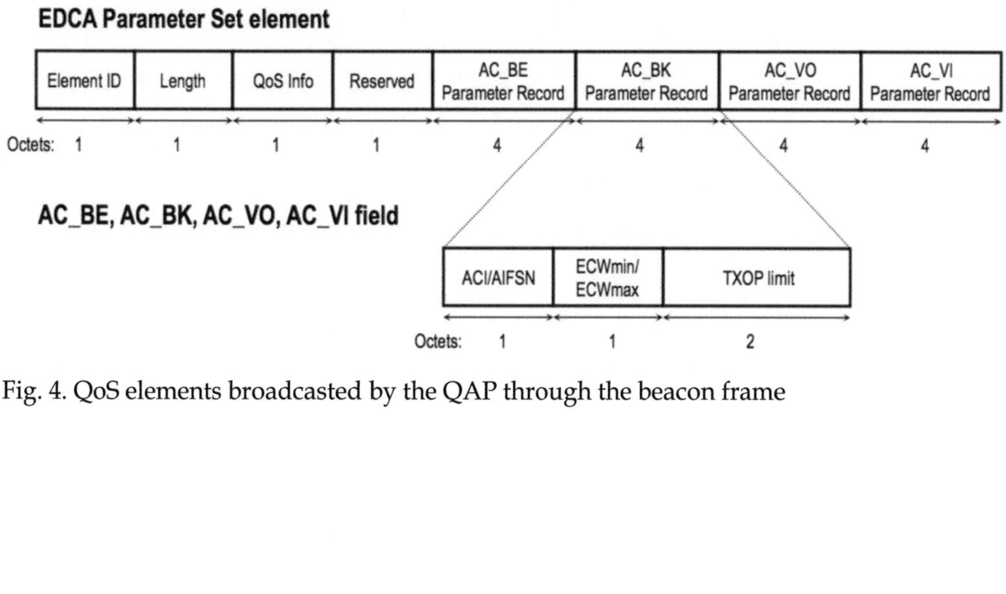


Fig. 4. QoS elements broadcasted by the QAP through the beacon frame

### 2.1 The fairness issue

Fairness between uplink and downlink channel allocation is a critical issue in infra-structured WLAN managed by a QAP. A QSTA must contend the uplink channel in order to transmit uplink frames towards the QAP. Downlink frames can be generated either by a wired network, i.e. by other stations connected to the Basic Service Set (BSS) through the QAP, or by other QSTAs inside the same BSS; in both cases the QAP acts as a concentrator and it must compete for channel access with all other QSTAs, before transmitting the frame.

In order to better understand the fairness problem, let us consider a WLAN composed by one BSS with $n$+1 stations, which include $n$ QSTAs and one QAP. Assuming, for simplicity, that all frames have the same length and equal priority, during a certain observing period $\Delta T$, the QAP, as well as each QSTA, gains an average channel time equal to $\Delta T/(n+1)$. In fact, all wireless stations have equal access probabilities, consequently the QAP receives the same treatment of a QSTA, in terms of channel allocation time. Each QSTA has to access the channel only for transmitting its own frames, while the QAP has to relay all downlink flows generated from wireless stations and addressed to other QSTAs. Hence, the QAP is penalized and downlink flows inevitably suffer performance degradation due to unfair channel allocation. In order to avoid this situation, the QAP should be assigned more resources than each wireless station.

In particular, we can consider two cases: a "balanced" and an "un-balanced" traffic scenario.

The balanced scenario is characterized by symmetric traffic conditions (i.e. VoIP conversation or peer-to-peer traffics), and downlink/uplink traffic volumes are similar. A fair allocation would assign to the QAP half of the channel allocation time; on the contrary, the QAP can exploit the channel only for $1/(n+1)$ of the time, for the reasons explained above.

In the un-balanced scenario, the wireless stations are more involved in downlink than in uplink direction; this reflects a common situation, in which the QSTAs are mainly involved in receiving web pages and audio/video streaming or in downloading files, while uplink traffic volume is more modest. Such a scenario would require at the QAP even more resources than those permitted by the EDCA scheme as it is (Grilo & Nunes, 2002).

Such an unfair behaviour increases collision probabilities between downlink frames, and, consequently, besides an inefficient resource allocation, the BSS is characterised by an increase of overhead, due to more back-off entities, and by a reduction of the BSS global throughput.

## 3. Regulation of EDCA parameters

Even if the EDCA scheme has improved QoS performance of multimedia traffics in WLANs, it is still possible to achieve even better results by suitably acting on its key parameters. In fact, a correct dimensioning of such parameters allows to reduce waiting times due to channel contentions, to achieve higher fairness between downlink/uplink allocation, and to decrease the number of collisions between frames belonging to the same AC and among different ACs.

In the following, we introduce how EDCA parameters can be tuned, so as to push the current 802.11e standard at its maximum performance, and we try to understand the limits beyond which new modifications to the standard algorithms are necessary.

The main advantages and disadvantages of the proposed modifications are also discussed.

### 3.1 Choice of AIFS duration

Equation (1) shows that the duration of the AIFS time interval for each AC depends on the choice of AIFSN, since slot time and SIFS duration are pre-determined values. There are several reasons which justify a modification of AIFSN with respect to the default values indicated by the standard (Table I). As it has been shown in (Thottan & Weigle, 2006), a reduction of AIFSN only for the ACs at the QAP permits to improve uplink/downlink fairness, because the QAP would be statistically advantaged during channel contention and, consequently, it has higher chances to obtain more channel time than the single QSTAs.

Moreover, a decrease of AIFSN for all stations (i.e., QAP and all QSTAs) leads to a reduction of waiting times and to an increase of WLAN global throughput. However, if AIFSN is reduced under a minimum threshold, several malfunctions could affect the 802.11e MAC.

In particular, when AIFSN is set to zero, from equation (1) we have that AIFS[AC] and SIFS coincide; so, if the random back-off algorithm selects a null CW value for a data frame, a collision can occur between that data frame and a higher priority frame, such as an ACK (Fig. 5).

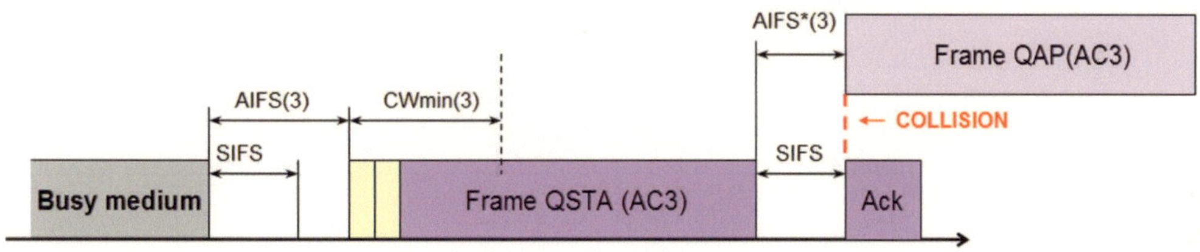


Fig. 5. Different choices of AIFSN, in relation to SIFS waiting time

A second situation, depicted in figure 6, shows the effect of setting AIFS[AC] lower than or equal to PCF Inter-Frame Space (PIFS), where PIFS represents the waiting time before entering in polling mode, as envisaged by HCCA. In such a case, the contention-free channel access mode defined by HCCA could be inhibited by EDCA data frames, and this fact cancels the reason for the existence of HCCA itself.

Moreover, when AIFS[AC] is set too low at the QAP, even if it respects the lower bounds described above, its CW would be dangerously shifted. In particular, it could happen that the AIFS related to an AC at the QAP is equal to or less than the AIFS of a higher priority AC at a QSTA. This fact is shown in figure 7, where AC2 at the QAP has higher transmission probability than AC3 at QSTAs, thus contradicting the original reason for prioritization. From the same figure, it is also evident that, when the AIFS[AC] value of a QSTA is set higher than AIFS[AC]+*CWmin*[AC] at the QAP, the QSTA frames risk starvation.

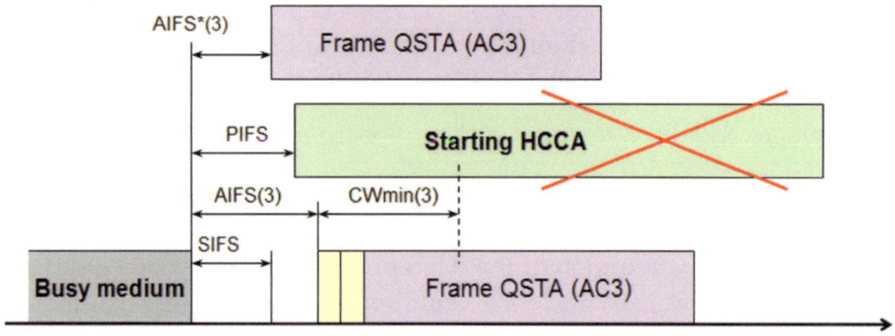


Fig. 6. Different choices of AIFSN, in relation to PIFS waiting time

A less hazardous and more interesting adjustment consists of dynamically increasing AIFSN for those ACs which are not characterised by real-time traffics, such as web or ftp, thus limiting their impact on a loaded network, as also stated in (Thottan & Weigle, 2006). Specifically, the work carried out in (Thottan & Weigle, 2006) investigates the performance of TCP traffic under different settings of EDCA parameters, by gradually adjusting AIFSN related to all the ACs of the QAP. Moreover, *CWmax* of AC1 at the QAP is halved down, in order to give more priority to the AP and to improve upstream/downstream fairness. However, in order to efficiently support QoS, it is better to operate jointly with AIFS duration, CW interval, and buffer sizes of the different ACs.

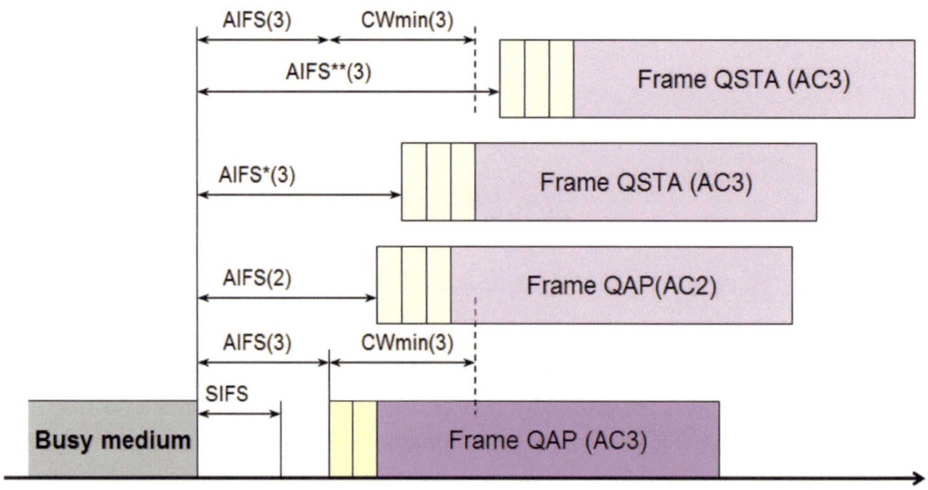


Fig. 7. Setting AIFS duration of different ACs

### 3.2 Choice of the contention window size

A careful tuning of the contention window parameters *CWmin* and *CWmax* can positively affect uplink/downlink fairness. In presence of light traffic conditions, low values contribute to reduce waiting times and to increase the global throughput. On the contrary, when traffic load becomes intensive, this choice could augment collision probability, because of the reduced number of slots inside a contention window (Nafaa, 2007).

(Wu et al., 2010) defines an *Adaptive Contention Window Adjustment* (ACA) scheme, which dynamically regulates the contention window size of each AC. In particular, it adjusts the CW size for each AC, according to the number of active connections for that specific AC.

Specifically, if the number of competing connections (inside the same AC) grows, the algorithm inflates the CW value, in order to reduce collision probability. On the other hand, if the number of contending stations decreases, it reduces the CW values, so as to limit frame delay.

However, this approach of changing CW size does not produce significant enhancements and, on the contrary, it introduces some drawbacks, such as a higher number of back-off entities. Moreover, an improper reduction of *CWmin* and *CWmax* for an AC at the QAP can result in assigning equal priority to a different AC of another QSTA. This happens when the BSS is characterized by many active stations and the QAP gets low values of CW parameters for improving fairness (fig. 8).

Another possibility envisages a proactive approach, by augmenting the CW size of an AC when the QAP realizes that the number of active traffic sources belonging to that AC is dangerously growing; this approach tries to prevent collisions and consequent performance degradation. This method can be joined with *TXOPlimit* adjustment.

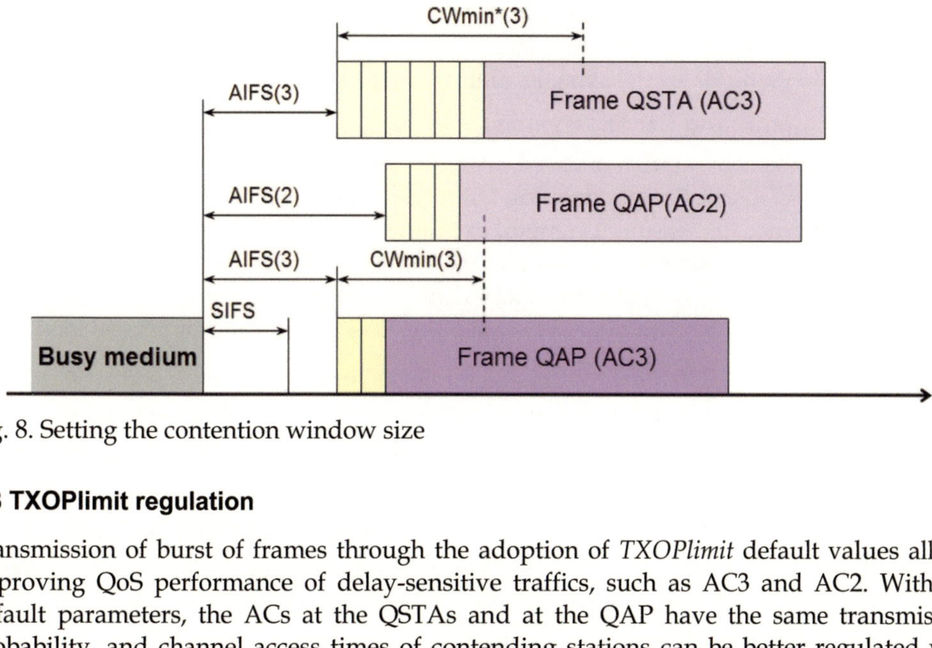


Fig. 8. Setting the contention window size

### 3.3 TXOPlimit regulation

Transmission of burst of frames through the adoption of *TXOPlimit* default values allows improving QoS performance of delay-sensitive traffics, such as AC3 and AC2. With the default parameters, the ACs at the QSTAs and at the QAP have the same transmission probability, and channel access times of contending stations can be better regulated with *TXOPlimit* option. Setting a unique value of this parameter for all stations (i.e., QSTA and QAP) can lead to unfair allocation between uplink and downlink traffics. In fact, the QAP has to deliver the whole downlink traffic to the QSTAs, but it contends the channel as one single entity. In order to give the QAP the possibility of transmitting all downlink frames, it is advisable to set a greater *TXOPlimit* value at the QAP with respect to QSTAs. Therefore, even under heavy traffic and with many contending stations, uplink/downlink fairness can be achieved through a careful dynamic management of the *TXOPlimit* parameter.

Of course, when an AC gains the channel, the duration of its transmission obviously influences the queue delay of the other ACs for all stations, and this can lead to frame starvation when there are many active stations. Long TXOP transmissions tend to reduce the number of channel contention opportunities and, consequently, all other stations have fewer chances to deliver their frames. As a result, it is evident that the regulation of *TXOPlimit* parameter must be carefully dimensioned.

In (Majkowski & Palacio, 2006), *TXOPlimit* is dynamically adapted on the basis of the number of frames allocated at the QAP (assuming that there is one queue for each AC). The average number of frames is considered rather than a simple measure of the actual length of AP's buffers, thus smoothing the effect of bursty arrivals. In order to avoid excessive channel occupation, TXOP maximum duration can be limited to the buffer size or to the value obtained when a certain collision probability is achieved, thus providing the optimum TXOP assignment for a given condition. This approach regulates the *TXOPlimit* value at the QAP, but it is not efficient under heavy loads in uplink direction.

In (Liu & Zhao, 2006), TXOP values are allocated in an efficient way for variable bit rate traffic with time varying profiles. TXOP is dynamically tuned according to the incoming frame size, with an estimation based on a variable bit rate video prediction algorithm, and to the current queue length. TXOP value is estimated as the sum of transmission time of the next incoming frame, of all frames in the transmission queue and of all the expected acknowledgements. The main drawback of this solution is computational complexity, introduced by the adoption of a Wavelet-Domain Predictor for the dynamic estimation of *TXOPlimit*.

In ETXOP (Ksentini et al., 2007), *TXOPlimit* is calculated each time the AC wins the contention, and the computation is based on AC's priority (inter-AC QoS) and on its flow data rate (intra-AC QoS). Each time a flow gains channel access, the algorithm checks the MAC queue in order to measure its length, to predict the number of frames and the mean frame size. Then, it computes at run time the most appropriate *TXOPlimit* that should satisfy QoS requirements of that flow. *TXOPlimit* is computed according to a distributed model which takes into account network availability, flow priority and its offered load

As the way to compute *TXOPlimit* is dependent on AC priority, for AC3 and AC2, ETXOP uses a value based on the number of frames in the corresponding MAC queue. Instead, for AC1 and AC0, ETXOP adopts static values, similarly to EDCA, aiming at limiting the number of frames sent in best effort and background bursts.

ETXOP offers more flexibility to network operators, by accommodating QoS requirements of network flows, regardless of their individual bit rates; this approach is more compliant with actual operated network practices (i.e., multi-bit rate environments).

In (Stoeckigt & Vu, 2010), authors discuss the implications of the TXOP parameter in terms of the maximum number of VoIP calls supported by IEEE 802.11 network. Here researchers give precedence to the QAP when competing for channel access, by assigning it a higher *TXOPLimit* value. However, such an increase causes the bottleneck to shift from the QAP to the wireless nodes, which have to wait an extended period of time before accessing the channel, thus increasing delays. Researchers have demonstrated that there is an optimal TXOP value, beyond which the WLAN voice capacity cannot be further improved. For this reason, an accurate recursive approximation formula has been studied, in order to calculate the achievable voice capacity in a WLAN for a given TXOP parameter. The processing of this approximation formula requires high computational resources. Moreover, the impact of the buffer size at the QAP on the number of obtainable voice calls is investigated, and an optimal buffer size has been defined for achieving the maximum voice capacity.

In (Feng et al., 2009), *TXOPlimit* is dynamically adjusted through a Random Early Detection (RED) mechanism, based on the queue length, which reflects the current network load. RED is a buffer management algorithm, whose packet drop probability linearly increases with the average queue length. According to this solution, the traffic load conditions are monitored at QAP and QSTAs queues. If the queue length is below a lower threshold, a smaller *TXOPlimit* value is adopted; if it overcomes this threshold, *TXOPlimit* increases linearly with the queue length. If the queue length is over an upper threshold, the maximum value of *TXOPlimit* is used. This algorithm is focused on QoS enhancement for video streams (AC2) and it is quite similar to (Majkowski & Palacio, 2006).

In (Andreadis & Zambon, 2007), a new algorithm is proposed, named *Dynamic-TXOP* (DTXOP), for the dynamic assignment of TXOP maximum duration. DTXOP is periodically updated according to the current traffic conditions of each specific AC, by computing the number of QSTAs involved in this AC and the amount of lost frames for each connection.

Unlike previous approaches, this algorithm is also able to reduce the QAP channel occupation time when uplink demand is greater than downlink demand.

Specifically, the proposed algorithm counts the number of lost frames of each AC in downlink/uplink directions, and the number of QSTAs demanding QoS requirements, during the *i-th* observing time interval (set equal to the beacon interval, 100ms). The term "lost frames" here refers to frames transmitted but not yet acknowledged by the destination station during the observing time. Lost frames are monitored because they are considered the main symptom of transmission problems, regardless of the events that caused such problems. Moreover, the number of QSTAs contending the channel is another key parameter involved in uplink/downlink fairness.

The described algorithm allows to enhance significantly QoS performance of multimedia traffics and to increase the WLAN global throughput. Although the major enhancements are denoted in real-time applications, such as video (AC2) and VoIP (AC3), TCP sessions such as web and file transfers (AC1 and AC0) are not significantly unfavoured, due to a global improvement in system's throughput and fairness.

### 3.4 Block ACK

In order to increase throughput and to reduce channel inefficiency introduced by ACK transmissions, a new acknowledgement scheme, called Block ACK, has been defined in IEEE 802.11e. This scheme allows transmitting consecutively multiple frames, inter-spaced by a SIFS interval. The acknowledgement of the transmitted block of frames is performed through a single aggregated ACK, called block ACK, thus avoiding the transmission of one ACK for each data frame.

The Block ACK frame is transmitted as a response to a control frame, called block ACK request frame (Fig. 9).

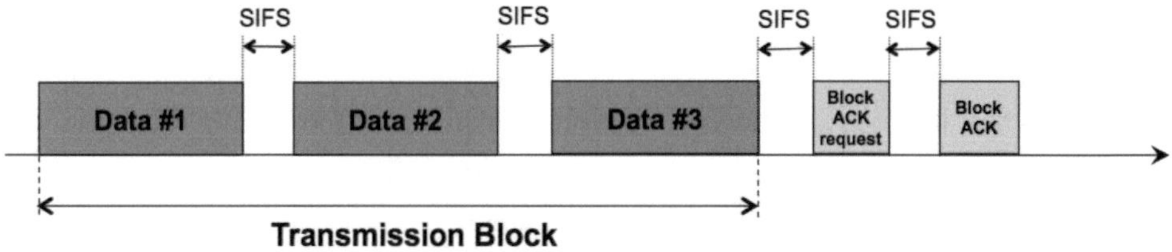


Fig. 9. Transmission of a block of frames and related Block ACK

Furthermore, the originator and the recipient have to set up this new acknowledgement policy, by exchanging an "add block ACK request" and an "add block ACK response" management frame. After this initialization, the frames that constitute the data block are transmitted and collectively acknowledged. The maximum number of data frames in a data block is specified during the initial setup phase, according to the buffer size at the receiver.

Since a data block can be composed of several data bursts, several TXOPs can be required for transmitting a data block (TXOP only determines the number of frames in a data burst).

The Block ACK scheme increases the aggregate throughput performance, due to a better efficiency of the MAC level acknowledgement mechanism. On the other hand, it increases delay, due to postponed acknowledgements. This delay is approximately proportional to the number of wireless stations contending the channel access.

Recent studies have been dedicated to the IEEE 802.11e block ACK scheme and to its performance evaluation.

In (Lee et al., 2010), a mathematical analysis of throughput and delay performance has been carried out, according to different channel access modes, channel errors (i.e. additive white Gaussian noise) and re-sequencing delay at the receiver. The accuracy of this model has been verified by comparing the numerical results with the ns-2 (NS2, 2011) simulation results. Also noisy environment were considered, showing that channel errors deteriorate throughput and delay performance due to retransmissions, and the adoption of *RequestToSend/ClearToSend* (RTS/CTS) technique over an error-prone wireless channel further degrades delay performance.

The work in (Yuan et al., 2004) introduces an adaptive block ACK scheme for infra-structured WLANs: high-rate hosts, experiencing good channel conditions, are given transmission priority by increasing the size of their data burst within a block-ACK.

In (T. Li et al., 2006), researchers show that an optimal block size, which maximizes the efficiency of block ACK mode, can be computed for each block transmission. The optimal block size is intuitively equal to the number of frames available in the transmission queue prior to delivery.

In (Wall & Khan, 2009), a novel distributed *Adaptive Block Size* scheme (ABS) is introduced. It dynamically adapts the block size on the basis of channel status and traffic characteristics, in order to achieve higher throughput and QoS efficiency. Specifically, the sender dynamically adapts the block size according to a specific function, assuming lower block size values when higher delays are experienced. Each block size adaptation has to be communicated between the sender and the recipient. This algorithm is focused on data delivery before lifetime expiry, and it provides protection from losses for real-time multimedia traffics. Being it a distributed scheme, ABS is applicable both to infra-structured and ad hoc WLANs.

## 4. Admission control schemes for WLANs

Under heavy traffic loads, 802.11e mechanisms for QoS support are no more sufficient and an admission control algorithm is needed, in order to avoid that the ingress of new traffic flows seriously damages the performance of active traffics. Several admission control algorithms were designed to play a central role at the aim to avoid network saturation and to protect QoS performance.

Furthermore, admission control algorithm is a key component to correctly manage QoS-based wireless networks, in order to adapt them to traffic load variations.

Specifically, in order to admit a new flow, it is important to satisfy two basic conditions:

- there are enough resources to meet the QoS demands of the new traffic;
- the upcoming traffic does not provoke a degradation of active traffics (i.e., bandwidth is efficiently exploited).

In this section we provide an overview of the state of the art on IEEE 802.11e admission control schemes proposed for preventing QoS degradation under heavy traffics.

Admission control algorithms can be distinguished in three categories, based on different criterions and methodologies:

- Measurement-based (threshold based and resource sharing based);
- Model-based;
- Measurement-aided, model-based.

### 4.1 Measurement-based admission control

In the measurement-based schemes, admission control decisions are supported by continuous measurements of network conditions, such as throughput and delay.

A threshold-based approach envisages that the QAP, and the QSTAs, if needed, measures the traffic conditions and the network status. Suitable upper or lower bound thresholds, delimiting the correct network load, are used to take the decision on the admission of new traffics.

The solution proposed in (Nor et al., 2006) introduces a *Network Utilization Characteristic* (NUC) of a new flow as a decision criterion for admission. NUC is defined as the fraction of channel utilization needed to transmit the flow over the network. If the total NUC, calculated as the sum over all the active flows and the new flow, is below a specific NUC threshold, the new flow is admitted.

This scheme is of easy implementation and can guarantee the QoS of high priority flows when the channel is heavily loaded, but it negatively affects the throughput of low priority flows to the detriment of fairness. Moreover, it appears critical the choice of NUC threshold values.

In order to enhance fairness among different QoS classes, an interesting solution could consist in the reservation of a minimum amount of resources for each class. This is the approach of resource sharing based admission control, as proposed in (Wu et al., 2010). Its idea is the following: the acceptance of a new traffic is performed only if QoS is preserved to a certain level which varies for different prioritized ACs. In other words, the QAP can accept a new AC admission request, only if both following criteria are satisfied:

- the average bandwidth requirements of all the existing AC traffics, with priority higher than or equal to the new flow, are preserved;
- the minimum reserved bandwidth of all existing traffics with priority lower than the new traffic is satisfied.

Similarly to the previous solution, (Xiao & H. Li, 2004) defines a fully distributed admission control algorithm, in which individual QSTAs accept or reject voice/video streams on the basis of local measurements.

During the beacon interval, the QAP calculates the amount of resources needed for transmitting all AC flows and it announces a transmission budget (i.e., the allowed channel time for each AC) via beacons sent to the QSTAs. According to the transmission count related to the previous beacon period and to the transmission budget announced by the QAP, each QSTA determines an internal transmission limit per AC, for each beacon interval. The local voice/video transmission time per beacon interval must not exceed the internal transmission limit per AC.

Also in (Kim et al., 2010) the admission control algorithm decides whether the flow can be admitted, basing on channel status and traffic information provided by the incoming flow. However, resource allocation to each AC is an extremely challenging issue, due to collisions among different ACs. For this reason, a priority access mechanism is introduced in order to carry out a correct channel contention among flows within the same AC, thus favouring high priority ACs and avoiding collisions between flows with different priority.

At each new admission request, average channel time usage ratio, average collision probability, and average back-off time are computed for each AC. These measurements allow the QAP to accurately predict the channel time usage for the admission of a new flow.

### 4.2 Model-based admission control

The model-based schemes build their performance metrics for evaluating the status of the network, on the basis of analytical models for the wireless system.

The Markov Chain Model is commonly adopted for IEEE 802.11 WLANs, but it has some limitations when EDCA parameters are introduced. For these reasons, researchers are working on new models for IEEE 802.11e (Chen et al., 2006).

In (Bellalta et al., 2007), an EDCA analytical model is adopted in order to estimate the minimum aggregated bandwidth required by all flows. By using this value and the maximum achievable data rate, the MAC parameters are adjusted on the basis of a set of predefined thresholds.

Although this solution takes in account all EDCA parameters and the uplink/downlink fairness issue, it remains difficult to define the correct threshold.

Furthermore, since analytical models are obtained on the basis of a few hypotheses to calculate QoS metrics of all flows, they do not accurately reflect the characteristics of real traffics and real channel time usage after a new flow has been admitted.

The study in (Cano & Bellalta, 2007) presents a model-based admission control scheme working jointly with a simple uplink/downlink fairness solution, which adopts different MAC parameters for the QAP and for the QSTAs. In particular, AIFS at the QAP is one unit lower than the value used at QSTAs, and its TXOP is increased proportionally to the number of downlink flows; furthermore, the *CWmin* value of uplink flows is set proportionally to the number of contending QSTAs.

When a new traffic needs to be activated, the admission control algorithm estimates if this operation preserves QoS for the existing flows. Estimation is performed by adopting the EDCA mathematical model developed in (Bellalta et al., 2007) to obtain the packet level performance of the system at both flow-level (blocking probability, average number of active flows, average bandwidth used etc.) and at packet level (flow throughput, packet delays, losses, etc.).

### 4.3 Measurement-aided, model-based admission control

A hybrid approach based on measurement-based and model-based schemes is arising as the main method for designing and implementing admission control algorithms.

The algorithm described in (Ksentini et al., 2007) is performed at each station and it takes into account measurements on network load, as well as the required data rate of the flow requesting for admission. Furthermore, this admission control scheme is able to predict the network conditions and hence, to estimate the achievable QoS performance, according to a channel model elaborated under the ETXOP assumption. Considering a specific AC, for each flow the QSTA starts estimating the maximum number of frames which can be inserted by that flow in the transmission queue, under the current network conditions. Afterwards, the QSTA checks whether this value satisfies the data rate required by the new flow. If the condition is verified, then the candidate flow is accepted, otherwise it is rejected.

Another scheme, described in (Zhu & Fapojuwo, 2007), estimates the achievable per-frame throughput and access delay and it verifies whether the ingress of new flows satisfies throughput and delay requirements. The new concept of *Virtual Service Interval* (VSI) is here introduced: VSI is calculated on the basis of network conditions and of the required access delays of flows with different priorities.

Performance metrics of throughput and delay are processed for each flow, on a frame-by-frame basis. These metrics are determined through a frame-based network model. In particular, the average service interval (i.e. channel occupation time) of admitted and applicant flows is calculated. The new flow is admitted if the average service interval is smaller than VSI.

This admission control method can be applied to both variable-bit-rate and constant-bit-rate traffics, since it considers per frame throughput and access delay as decision criteria.

The solution proposed in (Bensaou, 2009) is a threshold based admission control, according to which the QAP continuously monitors the channel and measures the contention probability. When a new flow requests admission, the authors adopt for this AC the non-saturation homogeneous equivalent model of DCF, in order to estimate the equivalent number of competing entities of the same AC. QAP computes the achievable bandwidth and expected delay of the new flow. If the bandwidth and delay requirements of new and admitted flows are all satisfied, then the new flow is admitted.

This solution is very simple, and more feasible if compared to other solutions, but the adoption of analytical model of a non-saturated IEEE 802.11 DCF introduces inaccuracy to the admission control in IEEE 802.11e EDCA scenarios.

The solution proposed in (Andreadis et al., 2008), based on a continuous monitoring of BSS channel resources, defines a resource sharing scheme for allocating the total bandwidth through resource reservation for each AC. For each new flow, the algorithm computes its resource utilization percentage by estimating the maximum achievable WLAN throughput. The key point of the proposed algorithm resides on the estimation of the channel global capacity, performed through the calculation of the average size of frames circulating in the network. In fact, the BSS global throughput is strictly correlated with the mean size of the frames generated by the transmitting stations (i.e., global throughput decreases with small frames) (Mangold et al., 2002). In (Andreadis et al., 2008), an experimental model for throughput has been proposed as a function of the average frame size. Based on this model, the estimation of the global throughput is constantly updated each time a new flow demands to be activated in the BSS.

In this way, a source with low throughput demands, but generating small frames, should be rejected, otherwise it would provoke a decrease of network throughput and of QoS performance of active traffics.

Furthermore, UDP flows (e.g., VoIP and video, usually mapped to AC3 and AC2) behave more aggressively with respect to TCP ones (e.g., ftp and web) and they are usually assigned a higher EDCA priority. For this reason, the admission control is restricted only to AC3 and AC2 classes, since AC0 and AC1 adapt themselves to the remaining bandwidth.

The activation of a new AC3 or AC2 traffic requires the following operations:

- evaluation of the frame size and of the required throughput for the new flow;
- re-computation of the current resource sharing percentages, checking whether the admission of the new flow would violate the sharing scheme;
- if the sharing scheme is not violated, the new AC flow is accepted, otherwise it is rejected.

The described algorithm is based on a low-collision and high-efficiency estimation model, so it works well in a scenario with increased uplink/downlink fairness. In fact, it is designed to work jointly with DTXOP algorithm (Andreadis & Zambon, 2007). The admission control scheme is able to exploit uplink/downlink fairness enhancements provided by DTXOP, in order to fine-tune the channel access control to the different ACs, without bandwidth waste or, on the other hand, QoS degradation of active flows.

## 5. Open issues

The main challenges regarding the enhancement of QoS performance for heterogeneous traffics in contention-based 802.11 networks are strictly linked to the dynamic adaptation of the *TXOPlimit* parameter and to the design of suitable admission control schemes.

Concerning TXOP, a key issue remains the way to correctly choose *TXOPlimit* value in order to preserve QoS of each flow, without penalizing network utilization of all competing flows. In fact, although large *TXOPlimit* values permit to increase performance of a specific traffic, on the other hand it causes high delays for other active traffics, with the risk of frame starvation and QoS degradation.

For this reason, finding the optimal transmission opportunity configuration (i.e. adopting recursive algorithms, thresholds, etc.) remains an open issue towards QoS management of heterogeneous traffics in a wireless network under different conditions.

Further network parameters could also be considered in the design of innovative *TXOPlimit* adaptation algorithms; for example, such algorithms can be calibrated for mobile or vehicular scenarios, by taking into account factors like channel noise, link quality, modulation, etc...

As regards the design of suitable admission control schemes, several issues have still to be faced. In particular, QoS performance should not be limited to the single wireless hotspot, but it should be preserved for the whole end-to-end path; consequently, methods to optimally map QoS requirements between different network layers have to be investigated, in order to dynamically adjust QoS on upper layers, while underlying network condition changes.

Furthermore, it is also interesting to explore how 802.11e techniques interact with applications and with higher-layer QoS schemes, and how Quality of Experience can be related to the Quality of Service concept.

Another possible field of investigation to achieve *TXOPlimit* regulation and admission control resides in the cross-layer approach and in game theoretical analysis methods.

Finally, the need to evaluate the trade-off between QoS techniques and energy consumption arises as a crucial point. For example, in EDCA, service differentiation and traffic prioritization can expand waiting times, thus affecting energy consumption. Since mobile devices are characterized by limited battery resources, these issues are very important in mobile networks and they require a very careful analysis.

## 6. Conclusion

The IEEE 802.11e standard is a concrete attempt in support of QoS, but this approach is not sufficient when the traffic volume increases. In this chapter we have provided an overview of the main techniques introduced to improve QoS performance in WLANs. A key concept for enhancing performance of IEEE 802.11e EDCA is the uplink/downlink fairness when allocating channel access time.

Firstly, the solutions related for fine-tuning EDCA parameters have been explored, mainly focusing on adaptation mechanisms of the *TXOPlimit* value, as a way to increase uplink/downlink fairness in resource allocation.

However, in presence of excessive traffic loads which lead to network saturation, admission control appears necessary for QoS preservation. For this reason, the major admission control algorithms for WLANs have been classified and discussed, as possible solutions for facing QoS degradation of active sources under heavy traffics. The joint adoption of fairness enhancement techniques and admission control schemes can increase efficiency and QoS performance.

We hope this work represents a valid contribution to clarify the state of the art about current studies on how to preserve QoS in contention-based (EDCA) IEEE 802.11e networks under heavy loads.

## 7. References

Andreadis, A. ; Benelli, G. & Zambon, R. (2008). An Admission Control Algorithm for QoS Provisioning in IEEE 802.11e EDCA, *Proceedings of the IEEE International Symposium on Wireless Pervasive Computing (ISWPC'08)*, pp. 298-302.

Andreadis, A. & Zambon, R. (2007). QoS Enhancement for Multimedia Traffics with Dynamic TXOPlimit in IEEE 802.11e, *The 3-rd ACM International Workshop on QoS and Security for Wireless and Mobile Networks (Q2SWinet'07)*, pp. 16-22.

Banchs, A. ; Azcorra, A. ; Garcia, C. & Cuevas, R. (2005). Applications and challenges of the 802.11e EDCA mechanism: An experimental study, *IEEE Network*, Vol. 19, No. 4, (July/August 2005), pp. 52-58.

Bellalta, B. ; Meo, M. & Oliver, M. (2007). VoIP Call Admission Control in WLANs in Presence of Elastic Traffic", *IEEE Journal of Communications Software and Systems*, Vol. 2, pp. 1-9.

Bensaou, B. ; Kong, Z. & Tsang, D.H.K. (2009). A measurement-assisted, model-based admission control algorithm for IEEE 802.11E. *Journal of Interconnection Networks*, Vol. 10, No. 4, (December 2009), pp. 303-320.

Cano, C. & Bellalta, B. (2007), Flow-Level Simulation of Call Ad- mission Control schemes in EDCA-based WLANs", *Proceedings of the 8th COST 290 Management Committee Meeting Universidad de Malaga*.

Chen, X. ; Zhai, H. ; Tian, X. & Fang, Y. (2006). Supporting QoS in IEEE 802.11e wireless LANs. *IEEE Transactions on Wireless Communications*, Vol. 5, No. 8, (August 2006), pp. 2217-2227.

Choi, S. ; Del Prado, J ; Sai Shankar, N. & Mangold, S. (2003). IEEE 802.11e Contention-Based Channel Access (EDCF) Performance Evaluation, *Proceedings of the IEEE International Conference on Communications (ICC'03)*, Vol 2, pp. 1151-1156.

Davcevski, M. & Janevski, T. (2005). Analysis of IEEE 802.11e QoS in Multimedia Environment, *Proceedings of IEEE International Conference on Telecommunications in Modern Satellite, Cable, and Broadcasting Services (TELKSIS'05)*, pp.45-48.

Feng, Z. ; Wen, G. ; Zou, Z. & Gao, F. (2009). RED-TXOP scheme for video transmission in IEEE802.11E EDCA WLAN, *Proceedings of the IEEE International Conference on Communications Technology and Applications (ICCTA'09)*, pp.371-375.

Grilo, A. & Nunes, M. (2002). Performance Evaluation of IEEE 802.11e, *Proceedings of the 13th IEEE International Symposium on Personal, Indoor and Mobile Radio Communications (PIMRC'02)*, Vol. 1, pp. 511-517.

IEEE. (1999). Part 11: Wireless LAN Medium Access Control (MAC) and Physical Layer (PHY) specifications, *IEEE Std*.

IEEE. (2005). Part 11: Wireless LAN Medium Access Control (MAC) and Physical Layer (PHY) specifications. Amendment 8: Medium Access Control (MAC) Quality of Service Enhancements, *IEEE Std*.

Kim, S. ; Cho, Y-J. & Kim, Y. K. (2010). Admission control scheme based on priority access for wireless LANs. *Computer Networks*, Vol. 54, No. 1, (January 2010), pp. 3-12.

Ksentini, A. ; Nafaa, A. ; Gueroui A. & Naimi, M. (2007). ETXOP: A resource allocation protocol for QoS-sensitive services provisioning in 802.11 networks. *ELSEVIER's Performance Evaluation (PEVA)*, Vol. 64, No. 5, June 2007, pp. 419-443.

Lee, H. ; Tinnirello, I. ; Yu, J. & Choi, S. (2010). A performance analysis of block ACK scheme for IEEE 802.11e networks. *Computer Networks*, Vol. 54, No. 14, (6 October 2010), pp. 2468-2481.

Li, T. ; Ni, Q. & Xiao, Y. (2006). Investigation of the block ACK scheme in wireless ad hoc networks. *Wireless Communications and Mobile Computing*, Vol. 6, No. 6, (September 2006), pp. 877-888.

Liu, H. & Zhao, Y. (2006). Adaptive EDCA Algorithm Using Video Prediction for Multimedia IEEE 802.11e WLAN, *Proceedings of the IEEE International Conference on Wireless and Mobile Communications (ICWMC'06)*, pp. 10.

Majkowski, J. & Palacio, F.C. (2006). Dynamic TXOP configuration for QoS enhancement in IEEE 802.11e wireless LAN, *Proceedings of the International Conference on Software, Telecommunications and Computer Networks (SoftCOM'06)*, pp. 66-70.

Mangold, S. ; Choi, S. ; May, P ; Klein, O. ; Hiertz, G. & Stibor, L. (2002). IEEE 802.11e Wireless LAN for Quality of Service, *Proceedings of European Wireless (EW'02)*.

Nafaa, A. (2007). Provisioning of Multimedia Services in 802.11-Based Networks: Facts and Challenges. *IEEE Wireless Communications*, Vol. 14, No. 5, (October 2007), pp. 106-112.

Nor, S. ; Mohd, A. & Cheow, C. (2006). An Admission Control Method for IEEE 802.11e. *Network Theory and Applications*.

NS2. The Network Simulator - ns-2. http://isi.edu/nsnam/ns/, Accessed 24 October 2011.

Stoeckigt, K. O. & Vu, H. L. (2010). VoIP Capacity—Analysis, Improvements, and Limits in IEEE 802.11 Wireless LAN. *IEEE Transaction on Vehicular Technology*, Vol. 59, No. 9, (November 2010), pp. 4553-4563.

Thottan, M. & Weigle, M.C. (2006). Impact of 802.11e EDCA on Mixed TCP-based Applications, *Proceedings of the International Wireless Internet Conference (WICON'06)*.

Wall, J. & Khan, J. Y. (2009). Efficient multimedia transmission using adaptive packet bursting for wireless LANs. *Computer Communications*, Vol. 32, No. 11, (3 July 2009), pp. 1271-1280.

Wu, H-T. ; Yang, M-H. & Ke, K-W. (2010). The design of QoS provisioning mechanisms for wireless networks, *Proceedings of 8th IEEE International Conference on Pervasive Computing and Communications Workshops (PERCOM'10 Workshops)*, pp.756-759.

Xiao, Y. & Li, H. (2004). Evaluation of distributed admission control for the IEEE 802.11e EDCA. *IEEE Communications Magazine*, Vol. 42, No. 9, (September 2004), pp. S20-S24.

Yuan, Y. ; Daqing, G. ; Arbaugh, W. & Jinyun, Z. (2004). High-performance MAC for highcapacity wireless LANs, *Proceedings of the 13th International Conference on Computer Communications and Networks (ICCCN'04)*, pp.167–172.

Zhu J. & Fapojuwo, A. O. (2007). A new call admission control method for providing desired throughput and delay performance in IEEE802.11e wireless LANs. *IEEE Transactions on Wireless Communications*, Vol.6, No.2, (February 2007), pp 701-709.

# 5

# QoS Adaptation for Realizing Interaction Between Virtual and Real Worlds Through Wireless LAN

Shinya Yamamoto[1], Naoki Shibata[2], Keiichi Yasumoto[3] and Minoru Ito[3]
[1]*Tokyo University of Science Yamaguchi,*
[2]*Shiga University,*
[3]*Nara Institute of Science and Technology*
*Japan*

## 1. Introduction

Recently, there are many studies regarding to MR (Mixed Reality) and AR (Augmented Reality)(ARtoolkit; Fudono et al. (2005); Ichikari et al. (2006)). Many research efforts have also been made for NVE (Networked Virtual Environment) and CSCW (Computer Supported Cooperative Work) (Fujimoto & Ishibashi (2005); Yang et al. (2010)). These technologies allow remote users to participate in social activities such as shopping, exhibition, sports, and game which are held in real space. However, VE requires high performance servers and high speed networks for supporting massively many users. AR and MR often require special devices (e.g. a motion capturing suit, an immersive display, a camera array, and so on) for capturing movements of users in a real world and reflecting them into a virtual space. In order to utilize VE and AR/MR for various purposes, a technology for realizing these techniques with low cost is essential. In our target application systems, the data amount exchanged between users is much bigger than that in typical network games, because we would like to capture and reflect more realistic and complex movement of objects for smooth interaction. However, we should notice that the available communication bandwidth in the Internet is limited. We introduce a DVE (Distributed Virtual Environment) mechanism to solve these problems. In order to realize the DVE mechanism, the following five criteria should be satisfied: (1) real and virtual users share a common virtual space; (2) users can freely change their positions and directions, and the changes are instantly reflected in other users' views; (3) each user can introduce objects into the shared space, and make actions, such as pushing and holding objects, and reaction should be reflected in other users' views; (4) the required apparatuses should not be special nor expensive; and (5) a massive number of objects can exist in the shared space.

There are some studies on communication architectures for MMOG (Hu & Liao (2004); Vleeschauwer et al. (2005); Yu & Vuong (2005)) and NVEs for remote cooperation (Chertov & Fahmy (2006); Eraslan et al. (2003)). Some of these existing studies realize scalability on sharing virtual space between many users using P2P technologies. However, they allow sharing virtual space and objects among only virtual users. On the other hand, the existing MR and AR technologies, the criteria (1) to (3) can be satisfied. However, they require special devices, servers and networks, and thus satisfying the criteria (4) and (5) is difficult.

In this chapter, we propose a framework named FAIRVIEW (Yamamoto et al. (2007)) which realizes smooth cooperation and interaction between real and virtual users satisfying all the criteria (1) to (5) using inexpensive devices off the shelf. To satisfy the criteria (1) to (3), FAIRVEIEW produces a hybrid shared space by overlapping a virtual space and a real space, and provides a mechanism for allowing the virtual and real users to observe each other. To satisfy the criterion (4), we suppose that virtual users use ordinary PCs with an internet connection, and that real users use wearable computers with HMDs (head mount display) or smartphones, with internet connection via wireless LAN. In FAIRVIEW, the information regarding to orientations and positions of real objects (called *AR information*, hereafter) are measured at short intervals using an existing AR measurement tool. The information is exchanged among user terminals, and objects are displayed as 3D graphics on the display of the virtual user's terminal. To achieve the criterion (5), we propose a mechanism for delivering AR information as well as action/reaction to the objects in real time (which we call *AR event delivery mechanism*, hereafter) to realize smooth cooperation among real and virtual users. Here, information of action/reaction consists of information of attribute (the color or the form, etc), action type(push, pull, etc), action strength and action direction. In addition, we define the *AR event* that consists of AR information and the information of action/reaction. The AR event delivery mechanism includes a QoS adaptation mechanism for controlling the intervals of transferring AR events between users so that the total transmission rates will not exceed the limit of available bandwidth. The adaptation mechanism decides the transfer intervals according to importance of each object for each user, which is determined automatically according to the distance and position of the object in the user's view.

To evaluate the proposed mechanism, we analyzed the required bandwidths and investigated the effectiveness of our QoS adaptation mechanism under some configurations with different numbers of objects. In addition, in the real space, users may not get sufficient efficacy of QoS adaptation because users share wireless bandwidth in a single AP coverage area. However, this problem can be improved by introducing multiple APs with appropriate AP selection. Therefore, we conducted computer simulation for measuring improvement of QoS adaptation with strategic AP selection when there are multiple APs in the real space. Furthermore, in the real-time interaction, users often do the sudden action, such as turn around. Therefore, to evaluate tolerance of our QoS adaptation mechanism to the sudden action of users, we measured the delay until reflecting update frequency of objects when the user's view changes, by using a prototype of FAIRVIEW. Simultaneously, to evaluate whether FAIRVIEW can achieve sufficiently short delays for real-time interaction, we measured the end-to-end AR event delivery delays through the AR event delivery mechanism in LAN. As a result, we confirmed that the proposed mechanism realizes smooth interaction involving a large number of real and virtual users/objects on an ordinary wireless LAN and internet environment.

## 2. Related work

Studies regarding to NVE and DVE, can be roughly categorized into two types: one is aiming at reducing traffic of game information and keeping scalability by efficiently managing Area of Interest (AOI) of users, and another is aiming at efficiently using computation and network resources.

In Hu & Liao (2004); Vleeschauwer et al. (2005); Yu & Vuong (2005), efficient AOI (Area of Interest) management methods are proposed. In Hu & Liao (2004), game space is dynamically divided based on Voronoi diagram for direct communication among players in a same fragment. In Vleeschauwer et al. (2005), the space is divided into small areas called micro cells. To distribute the load for processing events among multiple servers, the regions managed by each server are dynamically changed. In Yu & Vuong (2005), the shared virtual space is divided into honeycomb regions, and a mechanism based on Pastry (Rowstron & Druschel (2001)) is used to allow each user to receive information for players and objects in the player's AOI.

Although these existing NVEs enable efficient information exchange among virtual users, it is difficult to apply them to the mixed space of real and virtual worlds which requires a large amount of information exchange in real time on resource-limited wireless network, which is the target environment of FAIRVIEW.

In Chertov & Fahmy (2006); Eraslan et al. (2003), a load distribution and a QoS adaptation mechanisms for DVEs are proposed, respectively. In Chertov & Fahmy (2006), in order to cope with so called *area boundary problem* (inconsistency caused by neighboring areas managed by different servers), the whole shared space is managed by each of game servers, and game processing tasks are flexibly distributed among the servers. Unfortunately, this method supposes high performance servers and high-speed networks and treats only virtual users. In Eraslan et al. (2003), an IPv6-based network architecture called VESIR-6 is proposed for realizing a large-scale DVE where users can share a 3D virtual space and objects. Aiming at efficient utilization of network resources, VESIR-6 uses multicast for delivering object state updates to users, anycast for load distribution among servers, and IntServ/DiffServ-based QoS adaptation mechanism for regulating per-flow transmission rate. However, VESIR-6 does not suppose wireless network environment which is necessary for interaction between real and virtual users. Also, delivery of object state updates is managed only by joining/leaving the corresponding multicast group. Therefore, VESIR-6 cannot provide fine-grain QoS adaptation like the proposed method.

The most related study to our work is tele-immersion which captures the whole environment and reproduces it at geographically distant location. TEEVE (Yang et al. (2010)) displays 3D live visuals from each user's view in real-time, and constructs an environment for cooperative working. However, expensive and specialized devices and infrastructures such as 3D cameras and broadband network environments are needed to construct tele-immersion environment.

## 3. Overview of FAIRVIEW

This section overviews the functions of FAIRVIEW and presents example applications, then describes the target environment and the basic ideas for implementation.

### 3.1 Functions of FAIRVIEW

FAIRVIEW overlaps real space and virtual spaces, and provides the environment where users in real space (called *real users*) and users in virtual spaces (called *virtual users*) can interact as if they were in the same space. We call the overlapped space *hybrid space* [1] in Fig. 1. The main functions of FAIRVIEW are as follows: *Function (1) Providing the same view to real/virtual users:* Virtual users can change their positions in virtual space using keyboards and mice

[1] FAIRVIEW is also capable of overlapping more than one distant real spaces and provides the same view for the users in those spaces. For simplicity, we focus on interaction between the users in real space and virtual space.

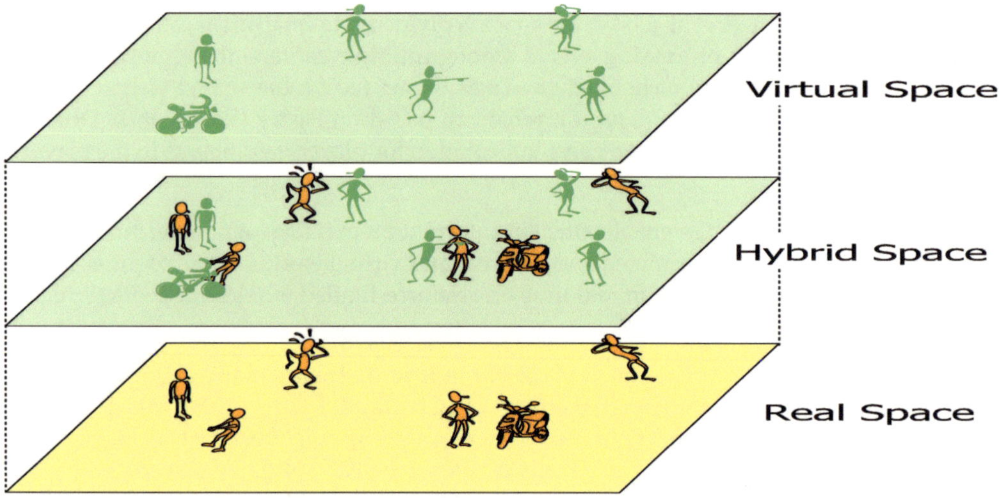


Fig. 1. Hybrid Space Produced by FAIRVIEW

like ordinary 3D first person shooting games. Real users can ordinarily move using their feet in real space. The users see objects in both virtual and real spaces according to their positions/directions. *Function (2) Voice conversation among users:* Users can talk with each other using their voice. The user's voice can be heard according to the user's position. This function can be realized with the technique in Yasumoto & Nahrstedt (2005). *Function (3) Object sharing:* Real objects and virtual objects can be registered in (and also unregistered from) the hybrid space. The registered objects can be seen by both virtual and real users. Users themselves are also objects[2]. Shared objects can be seen by all users whose views include the objects. We assume that 3D geometry data for the registered objects are prepared beforehand. *Function (4) Moving and bonding objects:* Users can move neighboring objects. Virtual objects can be moved by both real and virtual users, while real objects can be moved only by real users. Movement of objects can be observed by other users. Two or more objects can be bonded by specifying their relative positions. Bonded objects basically move together, but if a user moves a virtual part of bonded objects made of real and virtual objects, the bonded object is disengaged.

### 3.2 Applications of FAIRVIEW

We describe two example applications which are enriched by using the functions of FAIRVIEW described above.

*Virtual flea market* In this application, real and virtual users trade goods as if they were in a real market. Virtual users go shopping to a flea market in hybrid space. Real sellers register their goods to hybrid space. Virtual buyers see what kinds of goods are sold from distance. The sellers and buyers interact with each other via gestures and voice. The sellers show and explain details of their goods to virtual users by rotating and moving the goods. This kind of application includes exhibition, trade fair, and shopping center.

[2] Unregistered real objects can also be seen by real users, but in this chapter we assume that all objects are registered.

*Multiplayer game* In this application, virtual users participate in a paintball wargame held in real space. In paintball wargames, players possess airguns and targets, and shoot opponent groups' targets. Players whose targets are shot lose the game. Real players only need to register airgun and target. The virtual users cooperate with other real or virtual users. Virtual characters like a huge dinosaur can participate in the game. This kind of application includes attractions in theme parks, events on a street corner.

### 3.3 Devices and network used in FAIRVIEW

Table 1 shows the list of necessary equipments for the users. Real users use wireless small computing devices such as smartphones, and HMD with which the real view can be seen as the background of virtual view. These devices should not hinder real users' movements. The equipments also include positioning devices such as GPS receiver, sensors to detect position and direction, and audio input/output devices. We assume that inexpensive devices are available for these purposes. We use webcams and dedicated software like ARtoolkit (ARtoolkit) for positioning and detecting orientations of objects. Virtual users use ordinary PCs for FAIRVIEW.

| Type | Computer | Display | Network | Other |
|---|---|---|---|---|
| Real User | PC/Smartphone | HMD/etc | WiFi/etc | Sensor, Webcam |
| Virtual User | PC | LCD/etc | Internet | Mouse/etc |

Table 1. User Equipment

### 3.4 Basic ideas to implement FAIRVIEW

To realize an application using AR technology, we considered how the views seen by users are rendered inexpensively. In TEEVE (Yang et al. (2010)), images captured by 3D multi-cameras are processed and transferred through Internet2. Since we are aiming at inexpensively implementing the functions, we decided to measure AR information using inexpensive sensors. To realize this method, we have to resolve the following three problems: (i) measuring AR information accurately; (ii) transferring the measured AR information in real time; and (iii) rendering the objects.

For resolving the problem (i), we use GPS receiver if a user is in outdoor space. If a user is in indoor space, we use an indoor positioning method such as a method using wireless LAN access points (Kitasuka et al. (2003)), a method using speakers and microphones (Scott & Dragovic (2005)), Place Lab (Lab.), and Weavy (Kourogi & Kurata (2003)). The orientation of an object can be measured using rotational and translational acceleration sensors, methods based on image processing such as ARtoolkit (ARtoolkit) or the method in Fudono et al. (2005). To use the methods based on image processing, stickers with special patterns printed are put on objects, and the orientation and position of the objects can be obtained instantaneously from an image captured by a web camera. This kind of methods have cost advantage. In FAIRVIEW, AR information can be measured by combining the methods described above.

For the problem (iii), we assume that users have terminals capable of rendering 3D graphics. The problem (ii) is the main problem which we treat in this chapter. To resolve this problem, we need a delivery mechanism with which AR information can be exchanged among wireless and wired users in real time. We call this mechanism *AR event delivery mechanism*. In applications such as flea market or paintball wargame, hybrid space may have many objects and bandwidth shortage can occur as the amount of exchanged AR information becomes large. Therefore, we have to lower the frequency of delivering AR information. However, if the frequency becomes too low, the user experience of interaction may be ruined. Therefore, we propose a QoS adaptation mechanism which maximizes users' satisfaction by prioritizing transfer of AR information among objects based on relative position and orientation between users and objects. As a solution, we present the AR event delivery mechanism in Sect. 4, and the QoS adaptation mechanism in Sect. 5.

## 4. AR event delivery mechanism

| | notation | definition |
|---|---|---|
| Real Space | $R$ | |
| Virtual Space | $V$ | |
| Hybrid Space | $H$ | $H = (R, V)$ |
| Real Objects | $RO$ | $RO = \{ro_1, ..., ro_n\}$ |
| Virtual Objects | $VO$ | $VO = \{vo_1, ..., vo_m\}$ |
| Objects | $o$ | $o \in RO \cup VO$ |
| Real Users | $RU$ | $RU = \{ru_1, ..., ru_l\}$ |
| Virtual Users | $VU$ | $VU = \{vu_1, ..., vu_k\}$ |
| Users | $u$ | $u \in RU \cup VU$ |
| User Terminals | $node(u)$ | |
| Real User Terminals | $RN$ | $RN = \{node(u) \mid u \in RU\}$ |
| Virtual User Terminals | $VN$ | $VN = \{node(u) \mid u \in VU\}$ |
| All Object | $AO$ | $AO = RO \cup VO \cup RU \cup VU$ |
| Object including Users | $ao$ | $ao \in AO$ |
| AR Information | $AR(ao)$ | $AR(ao) = \{pos, angle\}$ |
| Attribute Information | $attribute$ | $attribute = \langle color, form, ...\rangle$ |
| Action | $action$ | $action = \langle type, direction, strength, ...\rangle$, $type \in \{push, pull, ...\}$ |
| AR Events | $ARe(ao)$ | $ARe(ao) = (AR(ao), attribute, action)$ |

Table 2. Definitions

This section describes details of the AR event delivery mechanism.

### 4.1 Notation

Let $R$ and $V$ be the target real space and the corresponding virtual space, respectively. Let $H = (R, V)$ be the hybrid space produced by overlapping $R$ and $V$. We suppose that $H$ is an axis-aligned rectangle on $x$-$y$ plane of 3D coordinate system. Let $hvec = (1, 0, 0)$ and $vvec = (0, 0, 1)$ be the horizontal standard vector and the vertical standard vector of $H$, respectively.

Let $RO = \{ro_1, ..., ro_n\}$, $VO = \{vo_1, ..., vo_m\}$, $RU = \{ru_1, ..., ru_l\}$ and $VU = \{vu_1, ..., vu_k\}$ be the set of real objects in $R$, the set of virtual objects in $V$, the set of real users (i.e., mobile users) and the set of virtual users (i.e., remote PC users), respectively. In addition, let $AO$ be the set of all objects including users, where each member object is denoted by $ao \in AO$.

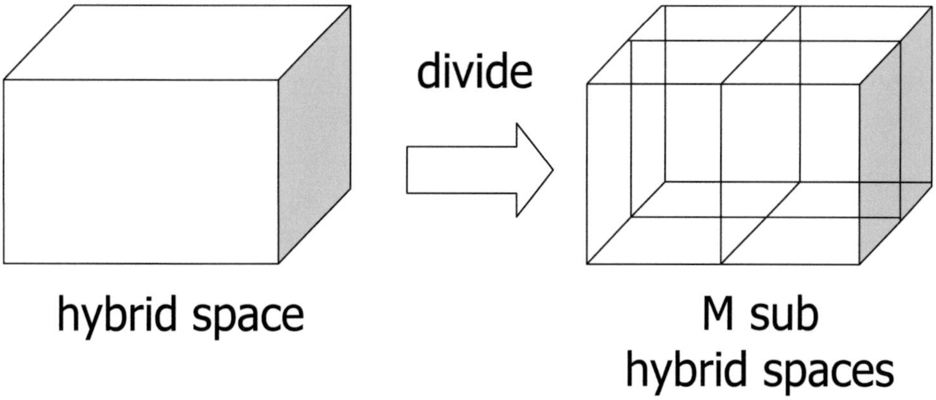


Fig. 2. Division of Shared Space

Note that $RU \subseteq RO$ and $VU \subseteq VO$. Let $node(u)$ be the user terminal of $u$ for each user $u \in RU \cup VU$. $RN = \{node(u)|u \in RU\}$ and $VN = \{node(u)|u \in VU\}$ are the real user terminals and the virtual user terminals, respectively.

For each object $ao \in AO$, let $AR(ao) = (pos, angleH, angleV)$ be the AR information of $ao$, where $pos$, $angleH$, and $angleV$ are its position on or beyond $H$, the horizontal angle to $hvec$, and the vertical angle to $vvec$, respectively. Each item of AR information is referred to by, e.g., $ao.pos$. In addition, let $ARe(ao) = (AR(ao), attribute, action)$ be the AE Event of $ao$, where $AR(ao)$ is the AR information of $ao$, $attribute = \langle color, form, ...\rangle$ is attribute information such as color and so forth, and $action = \langle type, direction, strength, ...\rangle$ is action to other objects, characterized by $type \in \{push, pull, ...\}$.

We suppose that each real user terminal in $RN$ can measure its user's AR information at a sufficiently high frame rate, e.g., 60 times per second, with equipment explained in Sect. 3. We also suppose that AR information of each real object $ro$ except for users can be measured by a user terminal in $ro$'s proximity with ARtoolkit and webcam.

### 4.2 Assumption on user communication

We suppose that the whole real space $R$ is covered by only one AP connected to the Internet. Therefore, each real user terminal in $RN$ can communicate with any virtual user terminal in $VN$.

Let $BW_{AP}$ be the available bandwidth between real user terminals and AP. Note that all real user terminals share the bandwidth. For each $vn \in VN$, let $bw_{AP}(vn)$ be the available bandwidth between $vn$ and AP. For each $vn \in VN$ and $rn \in RN$, let $bw(vn, rn)$ be the available bandwidth between $vn$ and $rn$. Note that $bw(vn, rn) = Min(BW_{AP}, bw_{AP}(vn))$. For each pair of virtual user nodes $vn1, vn2 \in VN$, let $bw(vn1, vn2)$ be the available bandwidth between $vn1$ and $vn2$.

### 4.3 AR event delivery mechanism

We choose some of the user terminals as server nodes to manage efficient AR events exchange. In order to reduce processing and traffic load of each server node based on user's AOI, we

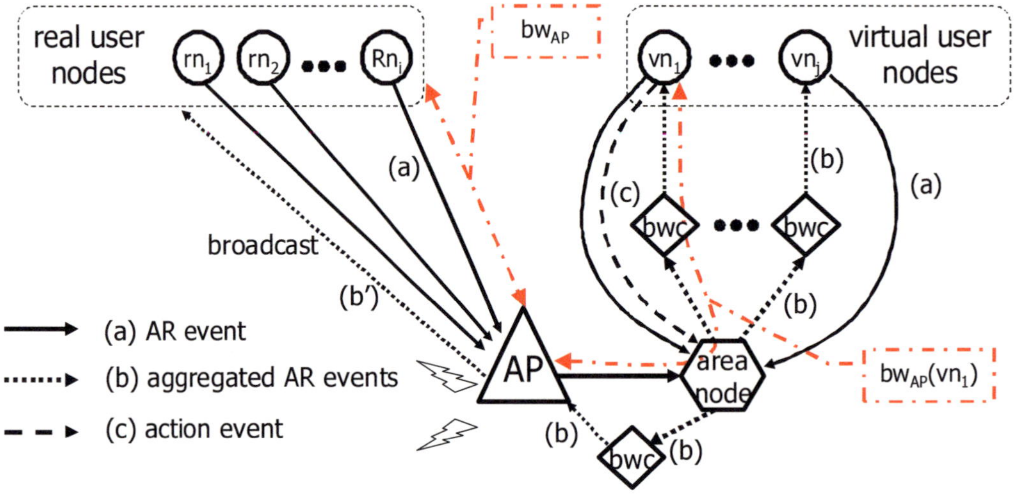


Fig. 3. Overlay Network for AR Event Delivery

divide the whole shared space $H$ into small rectangular sub-areas, as shown in Fig. 2, and assign a server node called *area node* to each sub-area. This is a similar approach to existing P2P-based MMOG gaming architectures such as Yamamoto et al. (2005). Let $an_A$ be the area node assigned to sub-area $A$.

The sub-area $an_A$ receives AR events of objects in $A$, and delivers the AR events to users (in $A$ and neighboring sub-areas) who are watching the objects.

In FAIRVIEW, since the network resource is tight due to wireless communication and calculating the reaction of an object as a consequence of an action is heavy for a real user terminal, we introduce new server node called *bandwidth controller node (bwc-node)*.

One bwc-node is prepared for each virtual user terminal $vn \in VN$ satisfying that $bw(vn, an_A)$ is less than necessary traffic for AR event transmission or for the set of all real user terminals[3]. Let $bn(u)$ be the bwc-node for user $u \in RU \cup VU$. Then $bn(u)$ applies QoS adaptation to the AR event stream between $an_A$ and $node(u)$, by monitoring $bw(bn(u), node(u))$.

Consequently, we construct the overlay network per sub-area, consisting of an area node, $N$ object nodes, $M$ user nodes, and $M$ bwc-nodes, as shown in Fig. 3, where $N$ and $M$ are the numbers of objects and users in the subarea, respectively. We will explain how these nodes exchange AR events below.

*User Node* In FAIRVIEW, (1) a user $u \in RU \cup VU$ can watch other objects in its view, (2) $u$ can be watched by other users since $u$ is also an object, and (3) $u$ can take an action to other objects.

For (1), $node(u)$ measures $u$'s AR information continuously and if the information differs from the last measurement, $node(u)$ sends the information as AR event to the area node $an_A$ (Fig. 3 (a)). For (2), $node(u)$ receives the AR events of the objects in $u$'s view and draws

[3] In FAIRVIEW, due to wireless bandwidth limitation, we restrict all real users to watching each virtual object at the same framerate.

the latest appearance of objects on display of $node(u)$. To receive AR event, $node(u)$ uses publish/subscribe model (Tanenbaum & Steen (2002)). Once $node(u)$ sends its AR event to the area node $an_A$, $an_A$ automatically identifies the objects in $u$'s view, and forwards the AR events of the objects to $node(u)$ via the bwc-node (Fig. 3 (b)(b')). As shown in Fig. 3 (b'), the real user terminals receive the AR events from the bwc-node assigned for real space $R$ via wireless AP by broadcast. For (3), when $u$ takes an action to virtual object $o$, $node(u)$ sends an action event containing power and direction, to $an_A$ (Fig. 3 (c)).

*Area Node* For each sub-area $A$ of the hybrid space $H$, a virtual user node is selected and assigned to the area node. A virtual node with sufficient network and computation resources is selected, e.g., by the lobby server when the application starts [4].

The area node $an_A$ manages the positions of the objects as well as the users' view in the sub-area $A$. When $an_A$ receives the AR events of all objects in $A$ from the corresponding user nodes and object nodes (Fig. 3 (c)), $an_A$ identifies the users who are watching part of other sub-areas neighboring $A$, based on their views calculated from their AR events (e.g., *pos* and *angleH*), and sends the AR events to the neighboring area nodes if needed. The sub-area $an_A$ also receives the AR events of such users in other sub-areas from the neighboring area nodes. Finally, $an_A$ sends the AR events of the objects in $A$ to the user nodes via the corresponding bwc-nodes.

*Bandwidth Controller Node* One bwc-node is assigned to each user node $u$, although one node may serve as the bwc-nodes of multiple users. The bwc-node $bn(u)$ monitors available bandwidth to its associated user node $node(u)$, and regulates transmission rates of AR event streams.

### 4.4 The overall flow

The overall flow is as follows.

**Step(1)** User Node $node(u)$ sends AR event $ARe(u)$ to Area Node $an$. Then, when User Node $node(u)$ has the managerial responsibility of an object $o$, it acquires the AR information of $o$, makes AR event $ARe(o)$, and sends $ARe(o)$ to $an$.

**Step(2)** Area Node $an$ adds an importance to AR event $ARe(ao)$ using the QoS adaptation mechanism described in Sect. 5.

**Step(3)** Area Node $an$ sends to Bandwidth Control Node $bn$, AR event $ARe(ao)$ with the importance.

**Step(4)** Bandwidth Control Node $bn$ regularly redistributes bandwidth with the QoS adaptation mechanism explained in Sect. 5 and decides update frequency of each AR event $ARe(ao)$.

**Step(5)** Bandwidth Control Node $bn$ sends the set of AR events to User Node $node(u)$ with the update frequency defined in Step(4).

**Step(6)** User Node $node(u)$ updates data of Hybrid Space $H$ with the set of received AR events.

**Step(7)** User Node $node(u)$ draws Hybrid Space $H$ with the updated data.

**Step(8)** returns to Step(1).

[4] The node selection/replacement algorithm for area nodes and bwc-nodes are omitted due to the space limitation. We use a similar technique to Yamamoto et al. (2005).

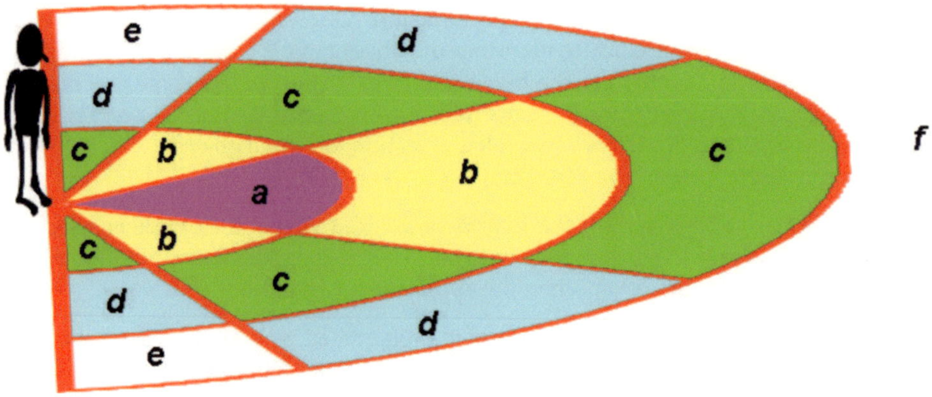


Fig. 4. User's View and Zones with Relative Importance

## 5. View-oriented QoS adaptation

The basic ideas of our QoS adaptation mechanism are as follows: (1) we decide relative *importance value* of each object according to how important the object is for a user; and (2) for each user, we regulate transmission rates of AR events of observable objects based on their importance values so that the sum of transmission rates are less than the available bandwidth.

### 5.1 Decision of importance value

Let $Watcher(o)$ be the set of users who can observe an object $o$ in $RO \cup VO$. The set of users $Watcher(o)$ is defined by

$$Watcher(o) \stackrel{def}{=} \begin{cases} \{u | u \in RO \cup VO, View(u) \in o.pos\} & (o \in VO) \\ \{u | u \in VU, View(u) \in o.pos\} & (o \in RO) \end{cases}$$

Where $View(u)$ is $u$'s view on hybrid space $H$, and represented by a half circle as shown in Fig. 4.

For each object $o$ in $RO \cup VO$ and each user $u$ in $Watcher(o)$, let $Imp(o, u)$ be the importance value of $o$ for $u$. Like in the real world, the importance value $Imp(o, u)$ should increase as the distance between $o$ and $u$ is shorter and $o$ is located nearer the center of $u$'s view. Therefore, we define each user's view as a half circle and divide it into 15 zones with five levels of importance $a, b, c, d$ and $e$, as shown in Fig. 4. Here, the objects located on zone $a$ have the highest importance, and the importance of objects on zones $b, c, d$ and $e$ decreases in this order.

### 5.2 Decision of framerates of AR events

The transmission rate of AR events for each object towards user $u$ is decided based on the ratio of its importance value to the sum of importance values of all objects in $u$'s view. The available bandwidth is distributed among the objects, and the framerate of AR events for each object is decided from the assigned transmission rate and the size of each AR event.

As mentioned in Sect. 4, AR events of each object are delivered to each user node through the corresponding bwc-node (see Fig. 3). The bwc-node drops packets of the AR events so that the transmission rate does not exceed the assigned bandwidth.

We describe our view-oriented QoS adaptation mechanism using examples. The QoS adaptation for AR event streams differs depending on the receiver type (i.e., virtual user or real user). Thus, we give two examples in the following subsections.

### 5.2.1 QoS adaptation for virtual user

Suppose that a virtual user $v \in VU$ is watching three objects $o_1, o_2$ and $o_3$ in his view. In this case, $node(v)$ receives the AR events of those three objects via the bwc-node $bn_v \in VN$ (see Fig. 3).

We assume that the available bandwidth between $bn_v$ and $node(v)$ is 1Mbps. We also assume that transmission rates for AR event streams of objects $o_1, o_2$ and $o_3$ are 0.5 Mbps respectively, that is, the sum of the streams is 1.5 Mbps. In this case, the available bandwidth (1 Mbps) is distributed according to the importance values of objects $o_1, o_2$ and $o_3$. Suppose that the importance values of objects $o_1, o_2$, and $o_3$ are 10, 25, and 15, respectively. As a result, portions of bandwidth $\frac{10 \times 1Mbps}{10+25+15} = 0.2Mbps$, $\frac{25 \times 1Mbps}{10+25+15} = 0.5Mbps$, and $\frac{15 \times 1Mbps}{10+25+15} = 0.3Mbps$ are assigned to the AR event streams, respectively. Based on this result, the bwc-node $bn_v$ controls transmission of the AR event stream of each object by dropping some of the received packets.

### 5.2.2 QoS adaptation for real user

Suppose that a real user $r \in RU$ is watching three virtual objects $o_1, o_2$ and $o_3$ in his view. In this case, $node(r)$ receives the AR events of those three objects via wireless AP and the bwc-node $bn_W \in VN$ (see Fig. 3).

In this case, we define the importance value $Imp_r(o)$ of an virtual object $o \in VO$ as the sum of the importance rated by the users observing the object $o$, i.e., $Imp_r(o) = \sum_{u \in Watcher(o)}(Imp(o, u))$. The framerate of AR events of each object is decided similarly to the case in Sect. 5.2.1, and the bwc-node $bn_W$ applies the QoS adaptation to the AR event streams to real users.

## 6. Experiments

In order to evaluate the proposed method, we conducted experiments on five items. First, we measured the required bandwidth for application of users and objects (Sect.6.2). Second, we measured the framerates at which users can watch the objects for two cases with and without our QoS adaptation mechanism (Sect.6.3). Thirdly, we measured the improvement of QoS adaptation with emulation in which users choose AP according to a strategy when there are multiple APs (Sect.6.4). Fourth, we measured the delay necessary for changing update frequency by QoS adaptation when the user's view changes, under some configurations with different numbers of objects using a prototype of FAIRVIEW (Sect.6.5). Finally, we measured the end-to-end delay through AR event delivery mechanism using a prototype of FAIRVIEW (Sect.6.5). Note that the experiments in Sect.6.2, 6.3 and 6.4 were emulations. The experiment in Sect.6.5 was conducted on a testbed using a prototype of FAIRVIEW with LAN.

### 6.1 Configurations

| | | Exp.6.2, 6.3 | Exp.6.4 | Exp.6.5 |
|---|---|---|---|---|
| User | Real | 100 | 100 | – |
| | Virtual | 100 | – | 1 |
| Field Size ($m^2$) | | 50×50 | 50×50 | 6×6– 120×120 |
| the number of APs | | 1 | 2, 4 | – |
| Bandwidth ($Mbps$) | wired | 10 | – | – |
| | wireless | 5 | 5 | 5 |
| Virtual Objects | | 50– 5000 | 500– 2500 | 8– 1683 |
| Packet Size | | 32 | 64 | 32– 64 |
| Max update frequency ($frame/sec$) | | 60 | 60 | 30 |

Table 3. Experimental Configurations

We show the experimental configurations in Table 3. The sizes of virtual space $V$ and real space $R$ are both 50 × 50 m. As user's view in Fig. 4, we set the radius of a half circle to be 15m and divided the half circle equally so that the angle and the radius of each zone are $1/3$ (i.e., $\pi/3$ and 5m), respectively. In this configuration, the average number of objects in each user's view is around 1– 2 when there are 50 objects in the whole space. In Fig. 4, zone $c$ exceeds in the number of objects compared with the zones $a, b$ because zone $c$ is the largest in the gross floor area. Therefore, in zones $a, b$, there is not much advantage than in zone $c$ when we give 2 times importance sequentially backward from zone $e$. Hence, we give zones $a, b$ importance of 4 times sequentially from $c$ backward. Thus, we assign the importance 64, 16, 4, 2, 1 to the zones $a, b, c, d, e$, respectively.

### 6.2 Required bandwidth for user terminals

The experimental configurations are shown in Table 3. The other setting is as follows. There is one wireless AP whose radio range covers the whole real space $R$. The available wireless bandwidth $BW_{AP}$ is shared among all real users. One area node and bwc-nodes are allocated on PCs on fixed wired network. 100 real users, 100 virtual users, and $n$ virtual objects are placed on the positions decided at random in the hybrid shared space, changing $n$ from 50 to 5000. The directions of the users are also set randomly. Each user terminal sends 60 packets of AR events every second, and the size of each packet is 32 bytes. The result is shown in Fig. 5. This result is average of 100 trials.

We measured the required network bandwidth between a user node and the corresponding bwc-node under the configurations in Sect. 6.1 for both cases with and without our QoS adaptation mechanism. The average results of 100 simulations are shown in Fig. 5.

Without the QoS adaptation mechanism, the required bandwidths for each real user terminal ("RU without QoS" in Fig. 5) and each virtual user terminal ("VU without QoS" in Fig. 5) exceeded the capacity (i.e., $10Mbps$ and $5Mbps$ ) when the number of objects in the whole space becomes more than 250 and 500, respectively. With our QoS adaptation mechanism, the required bandwidth is regulated below the capacity even if the number of objects increases to 5000 ("RU with QoS" and "VU with QoS" in Fig. 5).

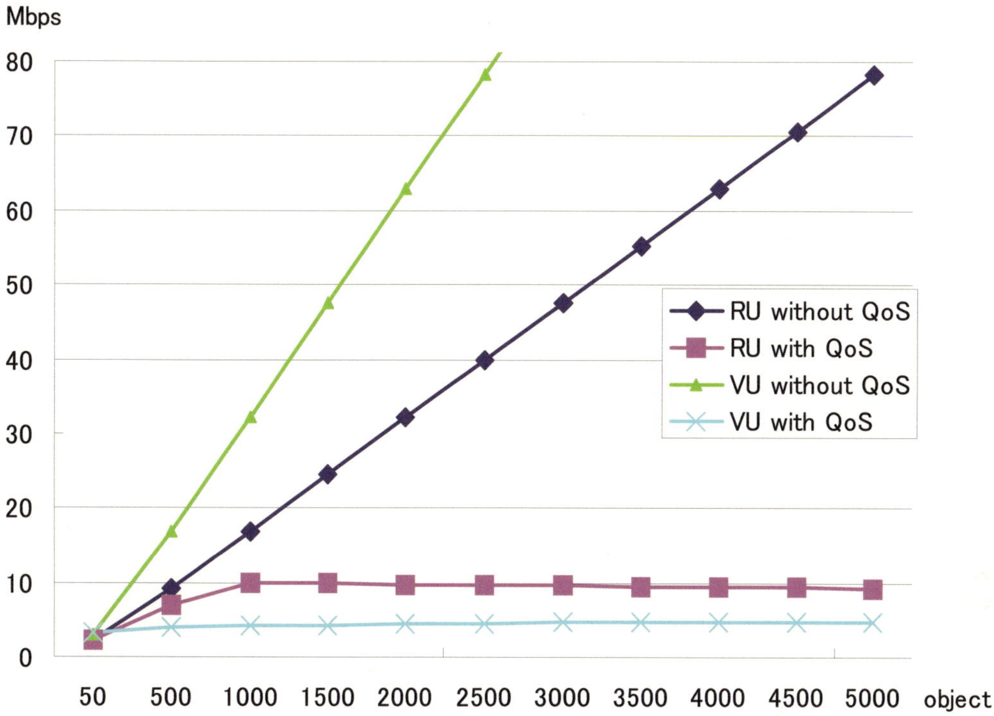


Fig. 5. Required Bandwidth for User Terminals

### 6.3 Impact of QoS adaptation

With our QoS adaptation method, AR events of more important objects are transmitted at higher transmission rates (i.e., framerates) than other objects, within the available bandwidth. To examine the effect of the QoS adaptation, we measured the framerates of AR events for objects in view zones *a* to *e* under the configurations in Sect. 6.2. The results are shown in Fig. 6 and Fig. 7.

In Fig. 6 and Fig. 7, the lines with labels $a, b, c, d$, and $e$ show the average framerates for objects on the corresponding zones, and the line with label *uniform* shows the average framerate when the bandwidth is uniformly distributed to objects.

Fig. 6 shows that each virtual user can watch important objects in zones *a* and *b* at higher framerates than *uniform*. Especially, framerate of objects in zone *a* keeps more than 50 frames/sec while the number of objects is less than 4500. The framerates of less important objects on zones $c, d$ and *e* are reduced below *uniform*.

Fig. 7 shows that each real user can also watch important objects in zones *a* at better framerates than *uniform*. However, the effect is smaller than the case for virtual users. The framerates of objects in other zones are reduced below *uniform*. This is because each object's importance value is decided as the maximum value among its watchers as explained in Sect. 5.1 and large portion of objects are regarded as important objects when the number of objects is large.

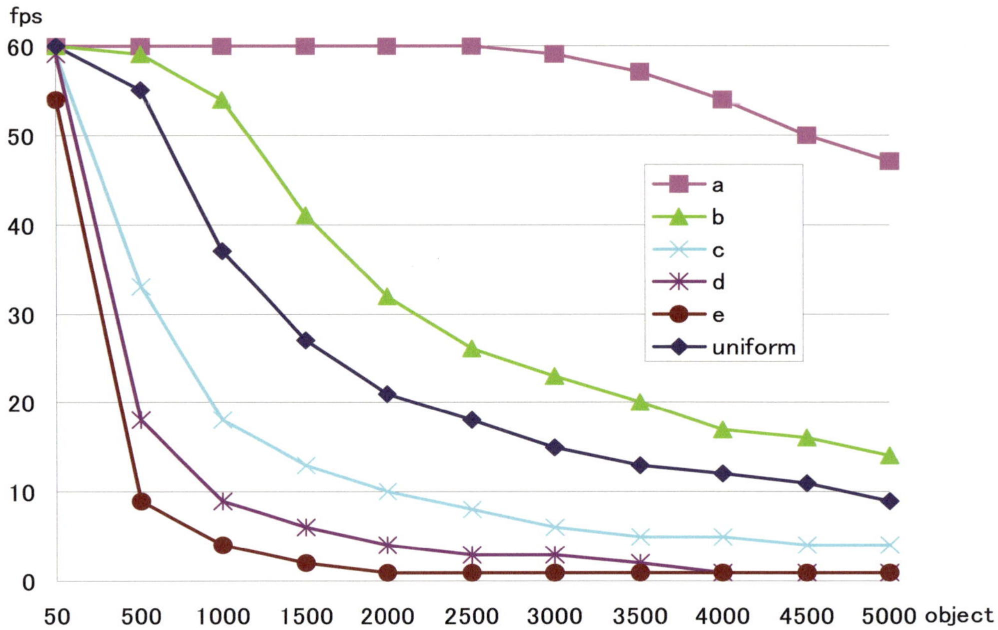


Fig. 6. Framerates Obtained by Virtual Users

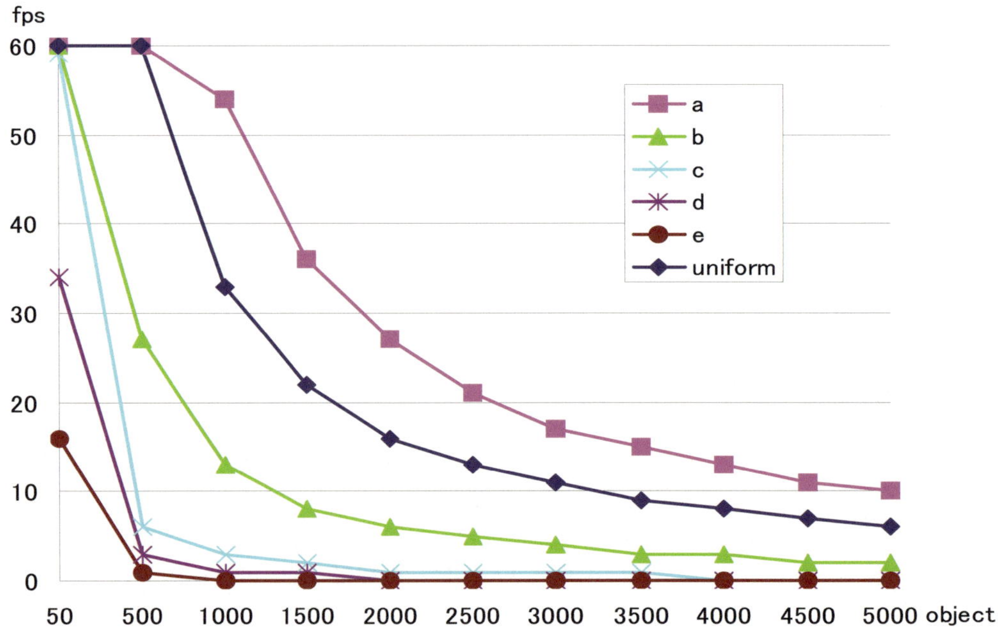


Fig. 7. Framerates Obtained by Real Users

However, the framerates of the objects in zone *a* are improved to a great extent when the number of objects is less than 2000.

### 6.4 Improvement of QoS adaptation control by AP selection

The real users share AP and receive AR events broadcast from AP. In the real space, each object's overall importance value is the sum of the values rated by all users sharing AP. Therefore, it cannot sufficiently reflect the importance of each user (shown as Sect. 5.2.2 and Sect. 6.3). If multiple APs are installed in real space, and users can choose AP which can reflect one's importance effectively, then we can improve the satisfaction of users. Therefore, we measured the improvement of QoS adaptation with simulation in which users choose AP according to a strategy when there are multiple APs. In this experiment, we installed 2 to 4 APs. In case the number of AP is 2, we equally divide real space in x-axis into two areas and placed one AP in each area. In case the number of AP is 4, we placed AP in each of the two by two matrix regions formed by equally sectoring a long x-axis and y-axis. In this experiment, to focus on the AP selection strategy, we supposed that a radio range is the whole area. In this experiment, the strategies that users are supposed to use are as follows.

1. UNIFORM: the user connects to AP with ID equal to $his/her\ ID$ mod $the\,number\,of\,APs$.
2. POSITION: the user connects to the nearest AP from his/her position.
3. VIEW: the user connects to the nearest AP from his/her position taking into account his/her view. Thus, comparing the his/her position and the end point of his/her view (e.g. distance = $15m$ and angle = 0 °), it employes the point whose distance to the AP is greater.

The results are shown in Fig. 8. This figure is typical boxplot that show how much frame-rate was assigned according to importance with strategic AP selection and the number of the AP. In this figure, the bar graph is Max-Min range and the box graph is the inter-quartile range. For example, in importance 64, 2 APs, "UNIFORM" in Fig. 8, we can see that there are the users for whom only 50 frames/sec was assigned to objects of importance 64, while many users can watch objects of importance 64 in 60 frames/sec. In this result, we can see how efficiently bandwidth allotment reflects the importance of the object for each user. In "UNIFORM" in Fig. 8, when several APs are available, the improvement is seen in comparison with the result in case the number of objects equals 2500 of Fig. 7. The value of framerates to be given disperse, because this strategy is random selection. In "POSITION" or "VIEW" in Fig. 8, dispersion of provided framerates for the importance of the user becomes small. In these results, "VIEW" was approximately equal to "POSITION". "VIEW" strategy removes users facing the other APs than the AP concerned. In spite of that, it is seen in Fig. 8, that bandwidth assignment in "VIEW" strategy seems more dispersive than "POSITION" strategy. The reason is that in "VIEW" strategy AP accepts far-off users looking towards itself as compensation does not accept far-off users looking outwards. As a result, the AP selection in consideration of the orientation of the user is effective in the QoS adaptation control for the real users. In QoS adaptation control for the real space, when there are some APs are available, we can get big benefits with strategic AP selection.

### 6.5 Processing delay of QoS adaptation

We measured the the coordination time of update frequency under some configurations when moving user's view by prototype system with different numbers of objects in LAN.

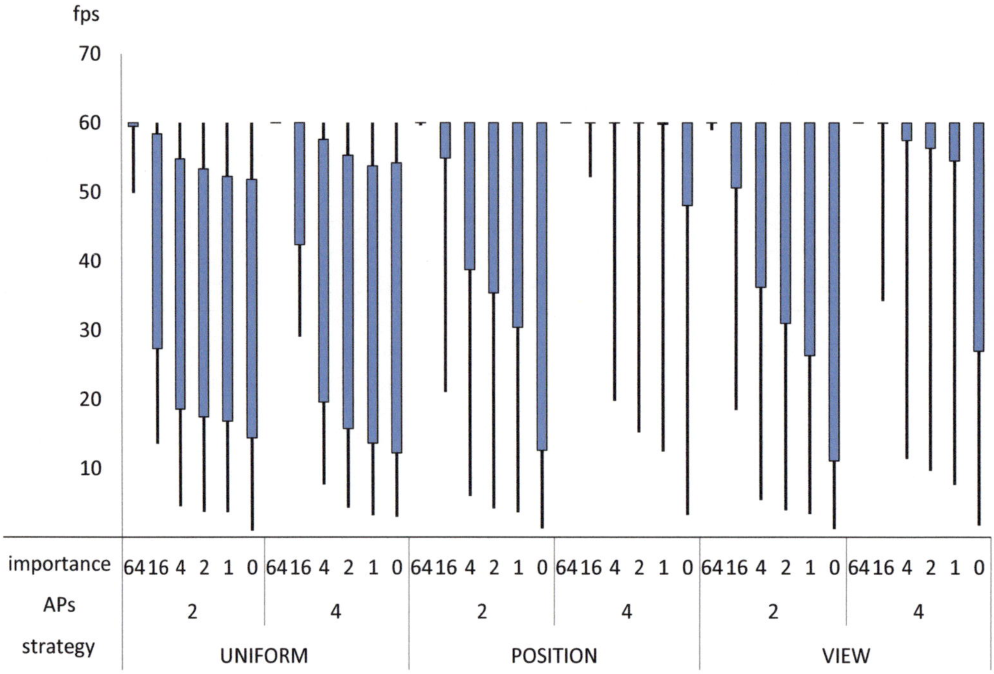


Fig. 8. Impact of AP Selection Strategy

Experimental configurations are shown in Table 3. The other setting is as follows. The user is placed in the center of the space. He/She walks at 1 $m/sec$ towards the destination decided at random. When he/she arrives at the destination, the next destination is decided at random again. In this experiment, we place $n$ virtual objects every 3 meters from the user. Then, the space has just only enough room for the all objects. Updates of the data of each object by the AR event, the transmission of packets, reception of packets, changes of the importance, and QoS adaptation control are preformed by parallel computation at about 30 times per second in the prototype for an experiment. Then, each process is asynchronous processed in an individual timeslot. The change of the importance is executed in a bwc-node and the QoS adaptation control is executed in an area node. We used TCP for communication protocol. The specifications of PCs are as follows: the area node and the bwc-node : CPU:Athlon64 X2 4200+, Memory:2GB, OS:Debian Linux (kernel 2.6.8); and a user node: CPU:Opteron 242(1.6 GHz)$\times$2, Memory:8GB, OS:Debian Linux (kernel 2.6.8). Programming Language is JAVA SE 1.6.0_05.

We measured, using the timestamp, the processing delay necessary for the recalculation of the importance and the recomputation of QoS adaptation control. In addition, we measured the end-to-end delay by the difference of timestamp from the time when user node sends AR event to the time when user node receives and updates data. These data do not include the delay of thread generation. The results are shown in Figs. 9 and 10. These results are average of 5 trials.

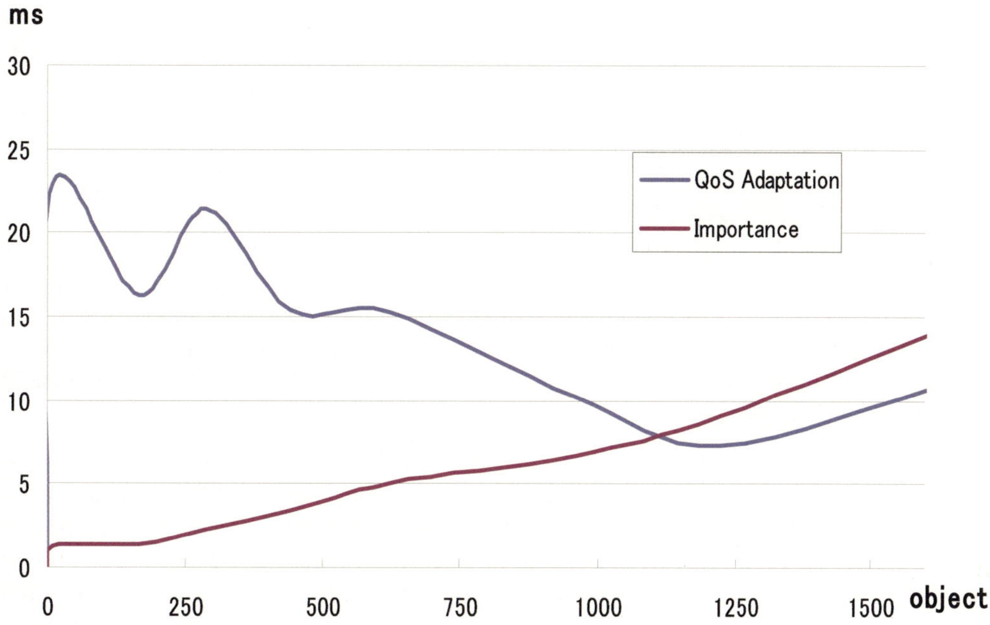


Fig. 9. Delay by QoS Adaptation

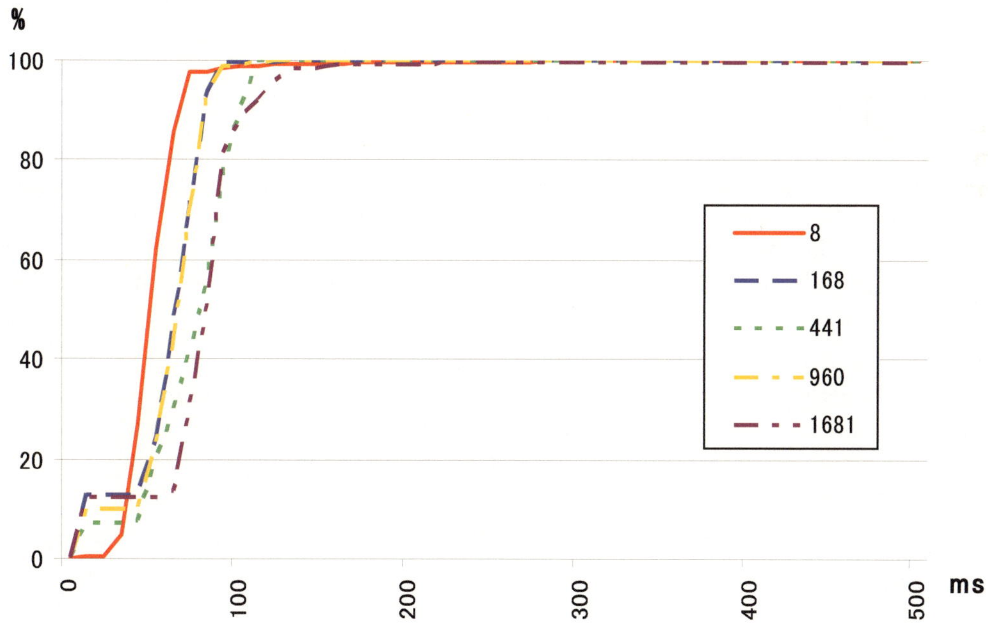


Fig. 10. Cumulated Distribution of End - End Delay

Fig. 9 is a graph of the processing delay of recomputation of the importance to change along the position of the avatar ("importance" in Fig.9) and recomputation by QoS adaptation control ("QoS Adaptation" in Fig.9). As a result, in the recalculation of the importance, a processing delay considerably grows as the number of objects increases because computational complexity increases according to the number of objects. The main cause of this processing delay is calculation of distance and angle of the avatar vs. the object. Therefore, when there are many objects, the importance is difficult to calculate in short time (e.g., as for every frame). As a solution, we can increase the number of area nodes or lower the frequency of the recalculation. The delay of QoS adaptation control changes by a timing of the processing. However, this delay stay lower than 25ms and does not depend on the number of objects. Furthermore, by the total of the recalculation of the importance and recomputation of QoS adaptation control, we can see the tolerance of user's sudden action (such as turns around), how much delay the influence is reflected on QoS adaptation when the importance of each object largely changes.

In Fig. 10, we show the end-to-end delay through AR event delivery mechanism including a result of Fig. 9 (Step(1)– (7) in Sect. 4.4). The end-to-end delay is quite satisfactorily and less than 200ms. The delay can be factored as follows: the recalculation of importance, the recomputation of QoS adaptation control, the wait time of the asynchronous processing. In Fig. 9, we see that the sum of the recomputation of QoS adaptation control and recalculation of importance is less than 50ms. Therefore, we show a main cause is wait time of the asynchronous processing. In addition, in Figs. 9 and 10, when there are objects more than 1,500, the total calculation delay is about 30ms and the influence is reflected within 200ms. If a human turns around and considers movement to adjust a focus to an object , we can permit this delay in practical uses. As a result, our framework can sufficiently cope with games requiring quick response.

## 7. Conclusion

In this chapter, we proposed a framework for interaction between real and virtual users in hybrid shared space called FAIRVIEW. In FAIRVIEW, the users share the information about moving objects including users themselves with Publish/Subscribe method on divided sharing space. In addition, in order to use FAIRVIEW in the ordinary environments, we proposed a QoS adaptation mechanism that can be implemented in a network with bandwidth limitation.

Through experiments, we confirmed that our method can handle hundreds of real and virtual users and thousands of objects with sufficiently fast framerate in an ordinary wireless LAN and internet environment. In addition, we confirmed FAIRVIEW could effectively offer QoS adaptation for real users by introducing strategic AP selection among a number of APs, compared with the case in which QoS adaptation is made through a single AP.

The future work is as follows. In the wireless communication environment, the delay tends to greatly fluctuate in comparison with the Internet. To such environments, we need to devise methods for realizing a short latency. In addition, e.g. in game applications, targets and obstacles are treated equally because our QoS adaptation control method treats all objects based on their values of importance. We will examine adjustment of the importance according to the kind of the object. Furthermore, we are planning to implement the proposed method as middleware library and evaluate its performance.

## 8. References

ARtoolkit . URL: *http://www.hitl.washington.edu/artoolkit/*

Chertov, R. & Fahmy, S. (2006). Optimistic load balancing in a distributed virtual environment, *ACM Network and Operating Systems Support for Digital Audio and Video (NOSSDAV '06)* pp. 74– 79.

Eraslan, M., Georganas, N. D., Gallardo, J. R. & Makrakis, D. (2003). A scalable network architecture for distributed virtual environments with dynamic qos over ipv6, *IEEE Symposium on Computers and Communications (ISCC '03)* pp. 10– 15.

Fudono, K., Sato., T. & Yokoya, N. (2005). Interactive 3-d modeling system using a hand-held video camera, *14th Scandinavian Conference on Image Analysis (SCIA2005)* pp. 1248–1258.

Fujimoto, M. & Ishibashi, Y. (2005). Packetization interval of haptic media in networked virtual environments, *4th ACM Workshop on Network and System Support for Games (NETGAMES '05)* .

Hu, S. Y. & Liao, G. M. (2004). Scalable peer-to-peer networked virtual environment, *ACM 3rd Workshop on Network and System Support for Games (NetGames '04)* pp. 129– 133.

Ichikari, R., Kawano, K., Kimura, A., Shibata, F. & Tamura, H. (2006). Mixed reality pre-visualization and camera-work authoring in filmmaking, *5th International Symposium on Mixed and Augmented Reality (ISMAR '06)* pp. 239– 240.

Intel Research Lab. URL: *http://www.placelab.org/*

Kitasuka, T., Nakanishi, T. & Fukuda, A. (2003). Wireless lan based indoor positioning system wips and its simulation, *IEEE Pacific Rim Conference on Communications, Computers and Signal Processing (PACRIM'03)* pp. 272– 275.

Kourogi, M. & Kurata, T. (2003). A method of personal positioning based on sensor data fusion of wearable camera and self-contained sensors, *IEEE Conference on Multisensor Fusion and Integration for Intelligent Systems (MFI2003)* pp. 287–292.

Rowstron, A. & Druschel, P. (2001). Pastry: Scalable, distributed object location and routing for large-scale peer-to-peer systems, *18th IFIP/ACM International Conference on Distributed Systems Platforms (Middleware)* pp. 329–350.

Scott, J. & Dragovic, B. (2005). Audio location: Accurate low-cost location sensing, *3rd International Conference on Pervasive Computing(Pervasive 2005)* pp. 1–18.

Tanenbaum, A. S. & Steen, M. V. (2002). *Distributed Systems - Principles and Paradigms*, Prentice Hall.

Vleeschauwer, B. D., Bossche, B. V. D., Verdickt, T., Turck, F. D., Dhoedt, B. & Demeester, P. (2005). Dynamic microcell assignment for massively multiplayer online gaming, *4th ACM Workshop on Network and System Support for Games (NETGAMES '05)* .

Yamamoto, S., Murata, Y., Yasumoto, K. & Ito, M. (2005). A distributed event delivery method with load balancing for mmorpg, *4th ACM Workshop on Network and System Support for Games (NETGAMES '05)* .

Yamamoto, S., Shibata, N., Yasumoto, K. & Ito, M. (2007). Qos adaptation for realizing interaction between virtual and real worlds in pervasive network environment, *ACM*

*Network and Operating Systems Support for Digital Audio and Video (NOSSDAV '07)* pp. 119–124.

Yang, Z., Wu, W., Nahrstedt, K., Kurillo, G. & Bajcsy, R. (2010). Enabling multi-party 3d tele-immersive environments with viewcast, *ACM Transactions on Multimedia Computing, Communications, and Applications (TOMCCAP)* 6(2): 1–30.

Yasumoto, K. & Nahrstedt, K. (2005). RAVITAS: Realistic voice chat framework for cooperative virtual spaces, *IEEE 2005 International Conference on Multimedia and Expo (ICME 2005)* pp. CD–ROM.

Yu, A. P. & Vuong, S. T. (2005). Mopar: A mobile peer-to-peer overlay architecture for interest management of massively multiplayer online games, *ACM Network and Operating System Support for Digital Audio and Video (NOSSDAV '05)* pp. 99–104.

# 6

# Custom CMOS Image Sensor with Multi-Channel High-Speed Readout Dedicated to WDM-SDM Indoor Optical Wireless LAN

Keiichiro Kagawa
*Shizuoka University*
*Japan*

## 1. Introduction

Visually-assisted optical indoor wireless local area network (LAN) is promising not only for high-speed but for offering intuitive user interface. Free-space optical communications (FSOC)(Jahns, 1994) are a key technology to create image-based ultra-fast wireless communication systems of the future. Compared with radio-frequency (RF) electromagnetic waves, the outstanding features of free-space light in terms of wireless communications are two-dimensional imaging with lenses as well as much higher frequency (> 100 THz), spatial and wavelength parallelism, and security. I believe that fusion of imaging and free-space optical communications can dramatically improve usability of indoor wireless LANs (Barry, 1994; O'Brien, 2005; Nonaka, 2006).

I have developed a new indoor optical wireless LAN system (Kagawa, 2003) that can offer a visually-intuitive user interface as well as high-speed data transfer based on two kinds of multiplexing. The key device is a complimentary-metal-oxide- semiconductor (CMOS) image sensor (Fossum, 1997). The device developed can receive several high-frequency amplitude-modulation optical signals at the same time as well as capture ordinary video-rate movies.

The image in the optical wireless LAN can be understood in two ways. One is that it shows the positions of communication nodes or the hub on the scene in a visual manner, which is very intuitive. For example, an actual scene image overlapped with identifiers like icons of communication nodes or the hub will add a feature of location awareness to the normally invisible network. Such interface combining the cyber space with the real will make the computer network more user-friendly. The other understanding of the image is spatial parallelism. Each light source out of the same line of sight is spatially separated on the image, which implies high-speed optical data acquisition by space-division-multiplexing (SDM) optical communications if the fast optical signals incident on different pixels can be read out concurrently.

In the FSOC, chromatic dispersive elements such as a grating are able to convert wavelength-division-multiplexing (WDM) to SDM by decomposing a single WDM light beam to multiple spots on the CMOS image sensor. The dedicated CMOS image sensor enables a fusion of imaging and FSOC, and can enhance the communication bandwidth of indoor wireless networks.

In this chapter, preliminary experiments of WDM optical data transmission with the dedicated CMOS image sensor are shown. In Sec. 2, fundamental configurations of the SDM-WDM indoor optical wireless LAN system and the communication modules are described. In Sec. 3, a wide-angle beam steering optics is mentioned. In Sec. 4, preliminary experiments with the dedicated CMOS image sensor and the beam steering optics are shown. Section 5 summarizes this Chapter.

## 2. SDM-WDM indoor optical wireless LAN

### 2.1 System configuration

Figure 1 shows a schematic drawing of the proposed space- and wavelength-division-multiplexing (SDM-WDM) indoor optical wireless LAN(Fujiuchi, 2004). The nodes connected to personal computers communicate each other via the hub installed on the ceiling. One of the most significant features of the proposed optical wireless LAN is that dedicated CMOS image sensors are utilized at both of the hub and the nodes as multi-point parallel photoreceivers as well as an image sensor for detecting the positions of communication target(s) on the scene.

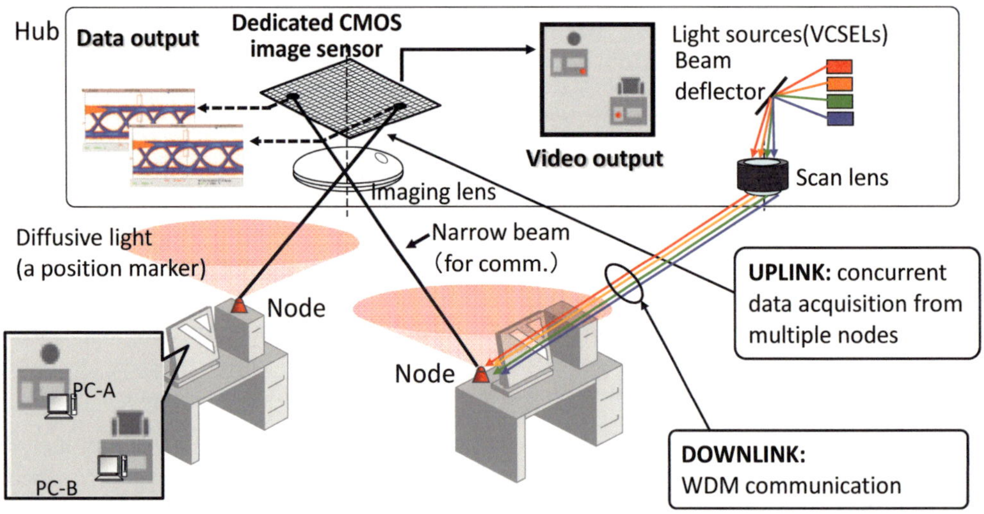


Fig. 1. SDM-WDM indoor optical wireless LAN using dedicated CMOS image sensors.

The dedicated CMOS image sensors are able to receive multiple optical signals at the different positions on them concurrently. With this ability, SDM can be adopted for the uplinks because the uplinks are multi-to-one connection. On the other hand, the downlinks are physically one-to-one connection between the hub and a node when the hub has only one transmitter. To increase the bandwidth of the downlink, parallelism of the dedicated CMOS image sensor is applied to implement a WDM feature.

The dedicated CMOS image sensor is equipped with two kinds of electric outputs: a movie of scene images and parallel digital output of optical signals. Each pixel operates in one of two functional modes: an image sensor mode and a photoreceiver mode. A communication link is established as follows; Firstly, the functional mode of every pixel is set to the image sensor mode, and each node and the hub emits a diffusive light with a predefined frequency or a sequence as a position marker. After the positions of the nodes and the hub are specified with some image processing, the operation mode of the pixels receiving the optical signals is set to the photoreceiver mode. Then, they begin to emit a narrow beam toward the detected counterpart(s). After the connection is established, the scene image captured at the hub is transferred to the node, and it is displayed with the identifiers of the nodes such as icons superimposed. The network users can recognize where the other communication nodes are in the communication area.

### 2.2 Communication modules

As shown in Fig. 2, the primary hardware components of the hub module are laser sources with different wavelengths, a wavelength multiplexer, a beam deflector, and a beam steering lens(Kagawa, 2008a) for the transmitter part, an imaging lens and the dedicated CMOS image sensor for the receiver part, and a position marker. To embody the optical multiplexer, dichroic mirrors or a blazed grating can be selected according to the size and the wavelength pitch of the laser diodes. The position marker is a diffusive light source such as a light emitting diode (LED) that illuminates the whole room, which is used as a position marker at the first step of the connection establishment. The composition of the node module is somewhat different. It has only a single laser source and no optical multiplexer in the transmitter part. On contrary, a blazed grating as a demultiplexer is added to the receiver part.

In Fig. 2, deflection of the light beam is introduced by the micro-electro-mechanical-systems (MEMS) mirror (Petersen, 1980; Miyajima, 2001). However, the amount of the deflection is not typically enough to cover the whole room. To overcome the limitation, a beam steering lens for amplifying the beam deflection was proposed (Kawakami, 2003).

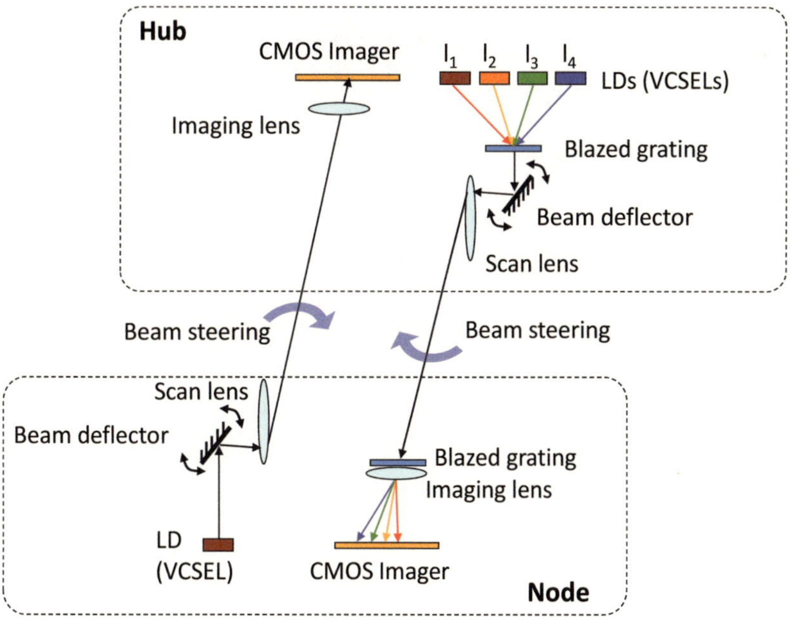


Fig. 2. Optical data transmission and beam deflection.

## 3. Dedicated CMOS image sensor

### 3.1 Sensor architecture

The CMOS image sensor dedicated to the proposed SDM-WDM indoor optical wireless LAN is a fusion of an ordinary image sensor and an array of photoreceiver circuits used in optical fiber communications. Figure 3 shows a block diagram of the dedicated CMOS image sensor. The pixel array is composed of $N_X \times N_Y$ pixels, and the sensor has $N_C$ parallel photoreceivers and data output channels. All signals for the communication are fully differential. For the image sensor mode, the image readout circuits are prepared at the top of the pixel array. The circuits below the pixel array are for the photoreceiver mode, which are composed of transimpedance amplifiers (TIA) with offset cancellation current sources controlled by automatic offset cancellers (AOCs) post amplifiers, bus drivers, an analog multiplexer to select $N_C$ outputs out of $N_X$ columns, bus receivers, a crosstalk reduction matrix operator, limiters to binarize the amplified photo-signals, and open-drain buffers.

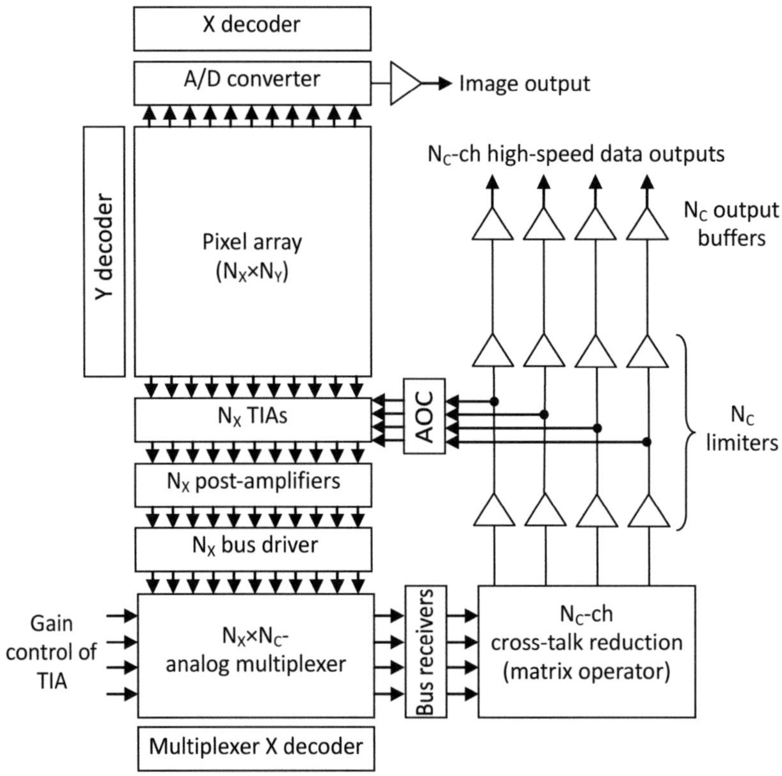


Fig. 3. Sensor architecture.

### 3.2 Pixel

#### 3.2.1 Circuits

To implement the multi-point optical data acquisition function, each pixel has digital control logic to change its behaviour according to the operation mode (Fig. 4). For the signal readout, pixel<i, j> has two kinds of output signal lines: one analog image output line (Vaps<i>) and two differential analog photocurrent output lines (Vsig<i-1>, Vref<i-1>, Vsig<i>, and Vref<i>) in the both sides of the pixel, which are selected by MEM_L<i, j> and MEM_R<i, j>. Note that <i, j> denotes the pixel position. xs<i> and ys<j> are horizontal and vertical addressing signals. rr<j> and rs<j> are row reset and select signals, respectively.

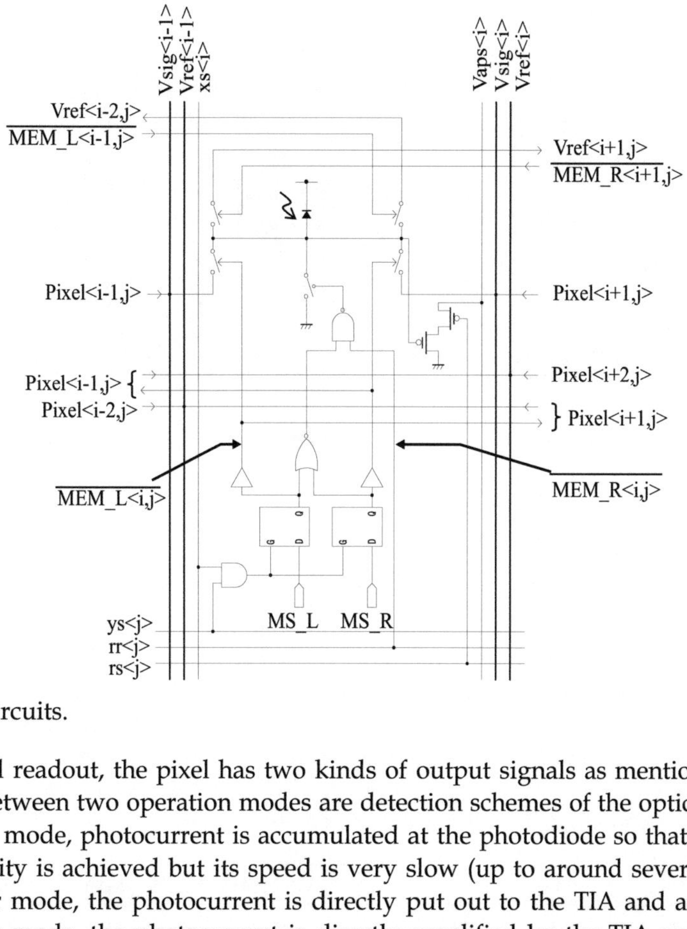


Fig. 4. Pixel circuits.

For the signal readout, the pixel has two kinds of output signals as mentioned above. The differences between two operation modes are detection schemes of the optical signal. In the image sensor mode, photocurrent is accumulated at the photodiode so that extremely high photosensitivity is achieved but its speed is very slow (up to around several MHz). In the photoreceiver mode, the photocurrent is directly put out to the TIA and amplified. In the photoreceiver mode, the photocurrent is directly amplified by the TIA prepared for each column without accumulation. Therefore, it can detect high-frequency signals but has low sensitivity. These complementary features are suited to detect the dim marker light in the imager mode and the strong narrow beam for communication in the photoreceiver mode.

### 3.2.2 Dynamic pixel reconfiguration for fully differential signaling

Another important feature of the pixel is dynamic differential reconfiguration to work as either a reference or a signal pixel. Without this feature, we cannot receive photo signals while capturing images because many kinds of noise sources on the CMOS image sensor and the print circuit board significantly decrease the signal integrity of the photoreciever circuits. While optical signals are amplified with capturing images, image readout digital circuitry becomes noise sources. Such common-mode noise can be suppressed by differential signalling. However, fully differential configuration needs replica pixels to put out the reference signal, which increases the pixel area (Zimmerman, 2003).

Figure 5 shows the schematic of the dynamic differential reconfiguration. In the figure, 2×2 pixels receiving optical signals work as a signal pixel, and four pixels in their both sides work as a reference pixel. With this method, differential signalling can be realized without replica pixels

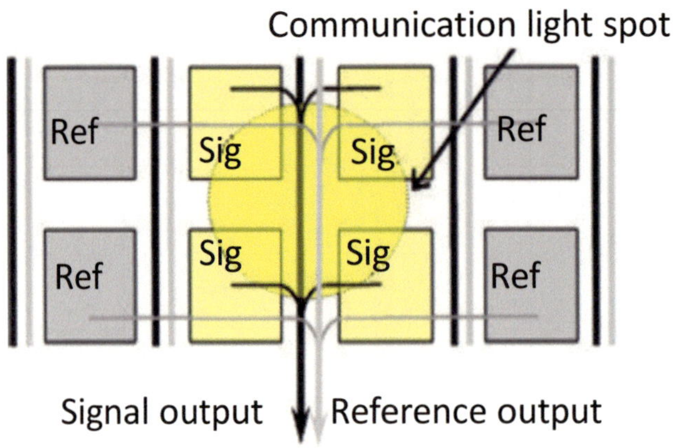


Fig. 5. Dynamic pixel reconfiguration.

### 3.3 Prototype sensor

Table 1 and Fig. 6 show specifications and a photograph of the dedicated CMOS image sensor. To build a demonstration system, photosensitivity of the CMOS image sensor should be high. Therefore, a high-sensitive but slow photodiode comprised of an N-well/ P-substrate-junction diode was selected. The bandwidth of the photo-amplifier was designed to be comparable with that of the photodiode. Figure 7 and Table 2 show a prototype board and specifications of an advanced version of the dedicated CMOS image sensor fabricated in a 0.18-µm CMOS technology. This chip has not been fully tested.

| | |
|---|---|
| Technology | 0.35-µm CMOS 2-poly, 3-metal |
| Chip size | 9.8 mm sq. |
| Pixel count | 64 x 64 |
| Pixel size | 100 µm sq. |
| Photodiode structure | N-well/ P-substrate |
| Fill factor | 16% (no microlens) |
| Photoamplifier | Regulated cascode amp (TIA)/ Cherry-Hooper amp (post amp and gain stage) |
| Total transimpedance gain | 2.5 kΩ - 2.5 MΩ (simulation) |
| Bandwidth | 7.9 MHz (simulation) |
| Number of channels | 4 |

Table 1. Specifications of the CMOS image sensor.

Fig. 6. Photograph of the prototype CMOS image sensor.

Fig. 7. Prototype board for CMOS image sensor in a 0.18-µm CMOS technology.

| | |
|---|---|
| Technology | 0.18-µm CMOS 1-poly, 5-metal |
| Pixel count | 180×84 |
| Pixel size | 31.25 µm×62.50 µm |
| Photodiode structure | Deep N-well/ P-well |
| Fill factor | 20% |
| Communication channels | 4 |
| Data rate | >1.0Gbps/ch (simulation) |

Table 2. Specifications of the CMOS image sensor in a 0.18-µm CMOS technology.

## 4. Wide-angle beam steering optics

### 4.1 Optical setup

The task of the optical transmitter is to deliver a narrow laser beam for communication to anywhere in the whole room with a size of more than 5 m by 5 m. For this purpose, a beam steering optics shown in Fig. 8 is proposed. The optical system is composed of three parts: a wavelength multiplexer, a beam deflector, and a beam-deflection-angle enhancer. The feature of the system is that the beam deflection enhancer is inserted at the exit of the beam steering optics to realize a compact and wide-angle optical transmitter.

To combine laser beams with wavelengths of λ1-λ4 into a single beam, the following equation should be satisfied. Note that δ, Δλ, d, and $f_1$ are the pitch and the wavelength difference of the adjacent lasers, the pitch of the blazed grating, and the focal length of lens, L1.

$$\delta = \left(\Delta\lambda / d\right) f_1 . \tag{1}$$

Because the availability of the MEMS mirror satisfying the requirements for the optical transmitter is not always good, the beam deflection part is implemented by a combination of a focusing lens (L2) on a compact two-dimensional stage and a collimator lens (L3). The focusing lens, L2, moves in the plane perpendicular to the optical axis. In this setup, an intermediate image is generated at the relayed image position in Fig. 8. When the displacement of L2 in one axis is represented by Δ, the output angle of the beam from L3, $\theta_{IN}$, is written by

$$\theta_{IN} = \mathrm{Tan}^{-1}\left(\Delta / f_3\right), \tag{2}$$

where $f_3$ means the focal length of L3. The beam steering lens as the beam deflection enhancer amplifies the input angle of the collimated beam, $\theta_{IN}$, with a certain angle gain to obtain the large output angle, $\theta_{OUT}$. With the beam steering lens, a wide range of the beam deflection angle can be achieved for a small displacement of L2.

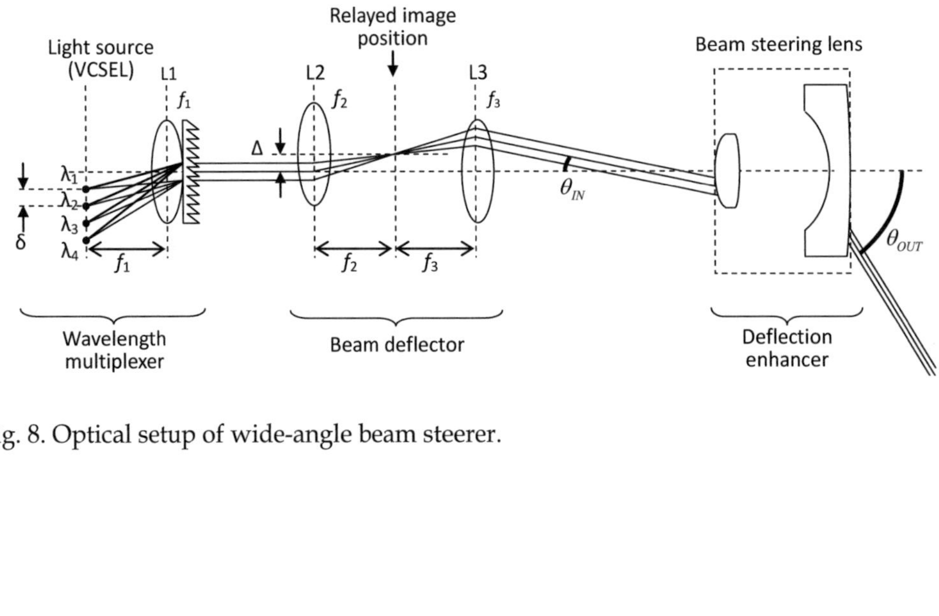


Fig. 8. Optical setup of wide-angle beam steerer.

### 4.2 Beam steering lens

The beam steering lens has a reverse-telephoto-type configuration (Smith, 2000). A prototype lens whose structure and specifications shown in Fig. 9 and Table 3, respectively, were designed and fabricated (Miyawaki, 2007; Kagawa, 2008a). The lens is designed for wavelengths of an 850-nm band, and its maximum field of view is 140-degree. The maximum gain of the beam angle is 3.5. A ray diagram is shown in Fig. 10. Because the left surface of the rear lens has a large curvature, the rear lens is composed of aspherical surfaces to minimize the output beam distortion. The feature of this beam steering lens is large tolerance of the alignment along the optical axis due to its infinite conjugate design, which makes assembly easier.

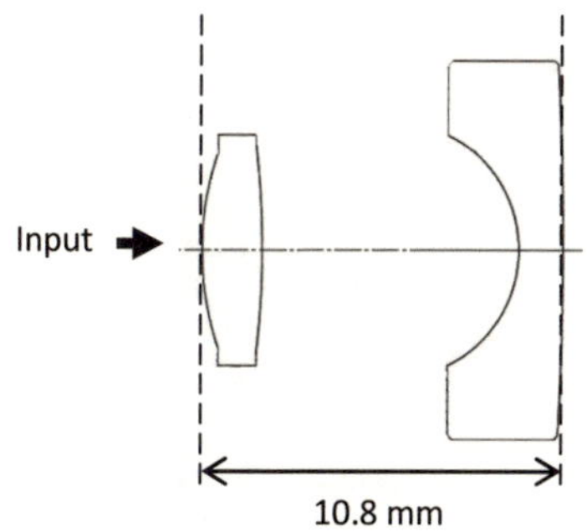


Fig. 9. Configuration of the beam steering lens.

| | |
|---|---|
| Wavelength | 845-851 nm |
| Focal length | 2.1 mm |
| Field of view | 140 degree (max) |
| Overall length after assembly | 10.2 mm |
| Working distance | 5.0 mm |
| Acceptable input beam diameter | 1.0 mm |
| Effective output diameter | 10 mm |
| Beam angle gain | 3.5 (max) |
| Material | K-SFLD6 (front),PBK40 (rear) |

Table 3. Specifications of the beam steering lens.

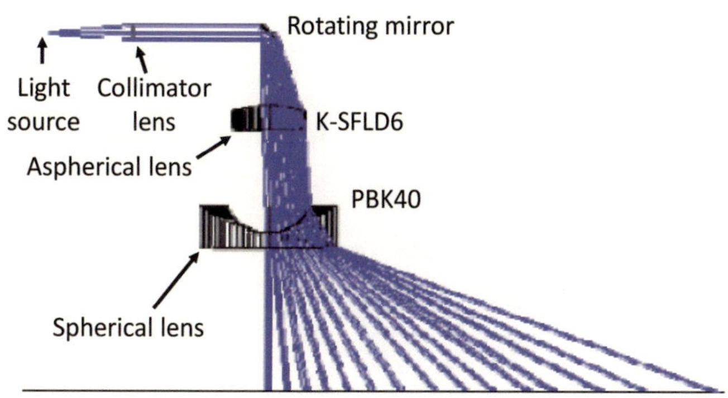


Fig. 10. Ray diagram.

### 4.3 Prototypes

A mounted lens is shown in Fig. 11. With this lens, the beam steering optics shown in Fig. 8 was constructed (Fig. 12). For simplicity, WDM was not implemented in this prototype. A laterally-single-mode GaAs vertical-cavity surface-emitting laser (VCSEL) was used as a laser source (FujiXerox, Model VCSEL-AS-0001, wavelength of 840-860 nm, maximum beam divergence of 20 degrees (FWHM), optical output power of 2 mW). The focusing lens, L2, was mounted on a movable compact stage (mechOnics, Model MS15, travel of 3.5 mm, maximum speed of 1.5 mm/s). $f_3$ was set to 2.2 mm. For this configuration, Δ should should range between about ±0.8 mm.

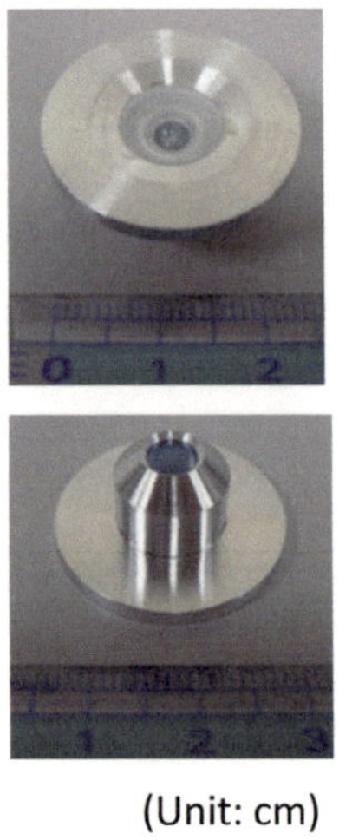


Fig. 11. Beam steering lens.

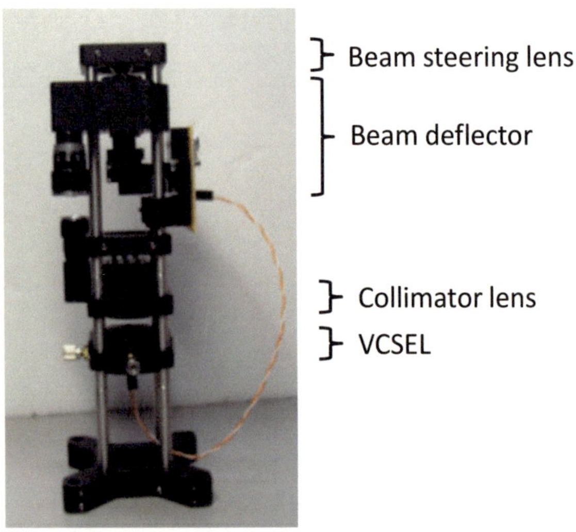


Fig. 12. Prototype of beam steerer.

## 5. Experiments

### 5.1 SDM data transfer

Figure 13 shows an experimental setup for uplink. The receiver of the hub was installed on the ceiling, and two transmitters ware placed on a desk. The vertical distance between the hub and the node was about 1.75 m. Transmitted data were generated by the data generator (Model DG2030, Tektronix, four digital output channels, data rate of 400Mbps per channel). For electric-to-optic (E/O) conversion, E/O converters (Model LL-650GI ($\lambda$=650 nm), LL-780GI ($\lambda$=780 nm), LL-900GI ($\lambda$=900nm), Graviton Inc., bandwidth of 1.2GHz) were used. The dedicated CMOS image sensor was controlled by a field-programmable gate array (FPGA). Control commands such as "start" or "stop image capturing", "set the pixels in the photoreceiver mode", and "set all the pixels in the image sensor mode" were issued by a personal computer through serial communication interface. An analog signal of the video output from the CMOS image sensor was converted to 10-bit digital value by an analog-to-digital conversion (ADC) chip. Then, it was transferred to the personal computer through a 10-bit parallel digital input-output (I/O).

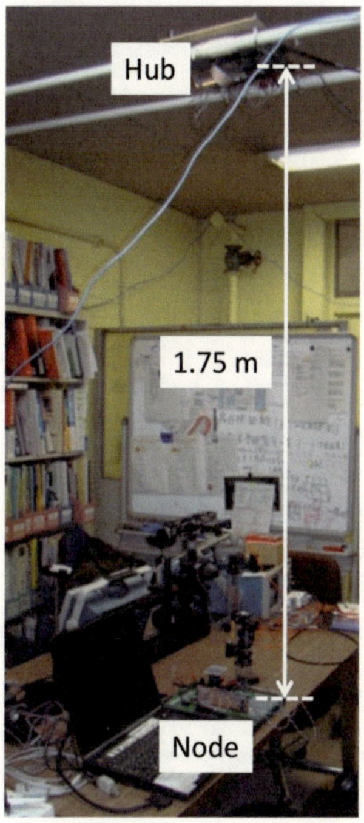


Fig. 13. Experimental setup for SDM data transfer.

Figure 14(a) shows a scene image captured at the hub when all pixels operated in the image sensor mode. A video lens (TECH SPEC, focal length of 4.3 mm, F/1.8) was used. Two regions indicated by the arrows in the figure are transmitters on the desk. Then, the operation mode of these pixels was set to the photoreceiver mode. Due to the dynamic reconfiguration, 4×2 pixels for each region became white. It is because these pixels are biased by the TIAs to amplify the photocurrent directly. The optical power before the imaging lens was about 200 μW.

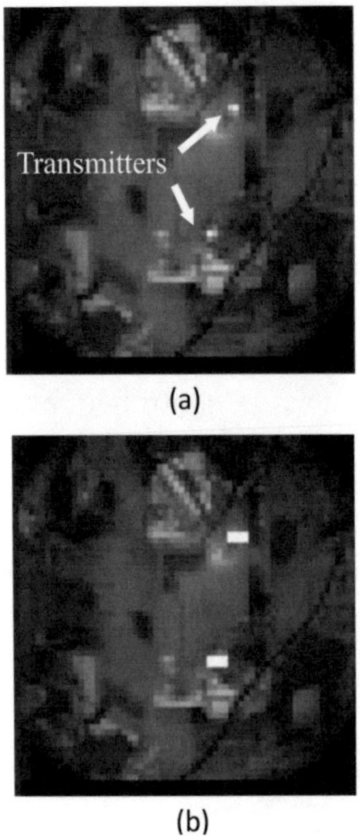


Fig. 14. Captured images for (a) all pixels in image sensor mode and (b) two rectangles in photoreceiver mode.

Figure 15 shows eye diagrams for the two nodes while images were being captured. The data rate was 10 Mbps/channel. This result shows that common-mode noise from the digital circuits on the CMOS image sensor to the photoreceiver circuits was well suppressed. The data rates were the same, but the wavelengths were different; for channel-1 and -2, the wavelengths were 780 nm and 650 nm, respectively. The delay becomes larger as the wavelength becomes longer. The differences of rise and fall times in the waveforms reflect dependency of the delay of the diffusion carriers generated in the silicon substrate of the CMOS image sensor on the wavelength. In the results shown in Ref. (Kagawa, 2008b) without differential signalling, the signal to crosstalk ratio (SCR), defined as the signal saturation level relative to peak-to-peak crosstalk, was 2.2 dB. With the full-differentiation technique, the SCR was about 11dB and 18dB for one of the differential pair and after subtraction, respectively.

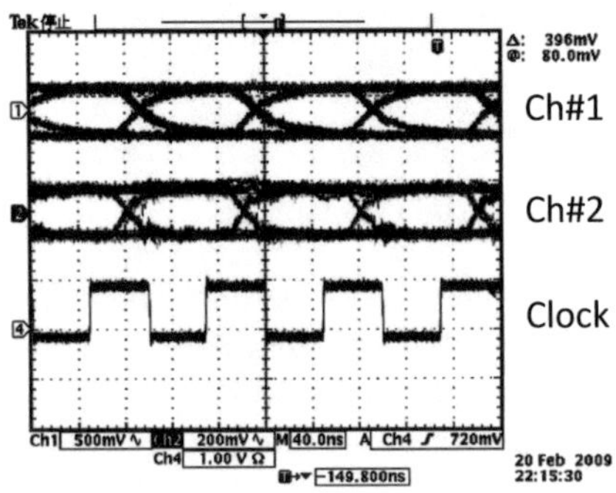


Fig. 15. Eye diagrams for the two nodes.

### 5.2 WDM data transfer

The experimental setup shown in Fig. 16 is for a WDM optical data transfer for the downlink. Three wavelengths were multiplexed by dichroic mirrors and travelled 1.65 m through the free space. At the receiver, the beam was demultiplexed to three optical spots on the CMOS image sensor through a transparent blazed grating (grating constant of 300/mm, diffraction efficiency of more than 60%) and a imaging lens (SCHNEIDER, wavelength range of visible to infrared light, focal length of 12 mm, F/1.4). The frequency of each wavelength channel was set to be different. The waveforms for three wavelengths are shown in Fig. 17. The result shows that demultiplexing of the multiplexed optical signals was successfully achieved.

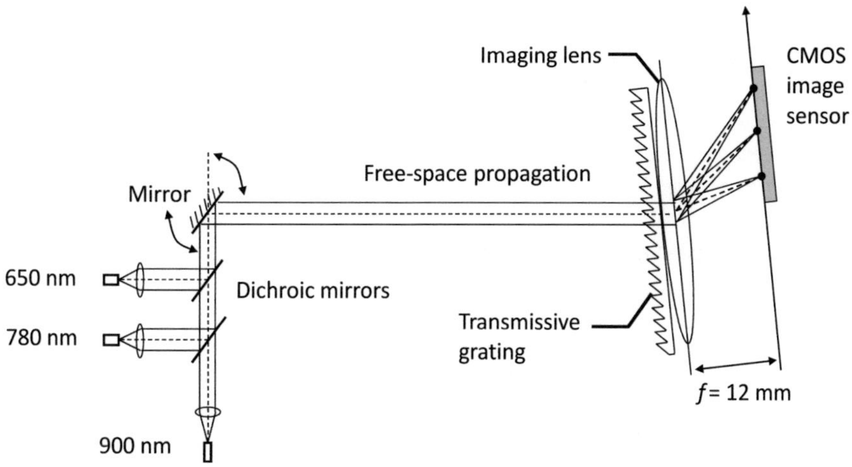


Fig. 16. Optical setup for WDM optical data transfer.

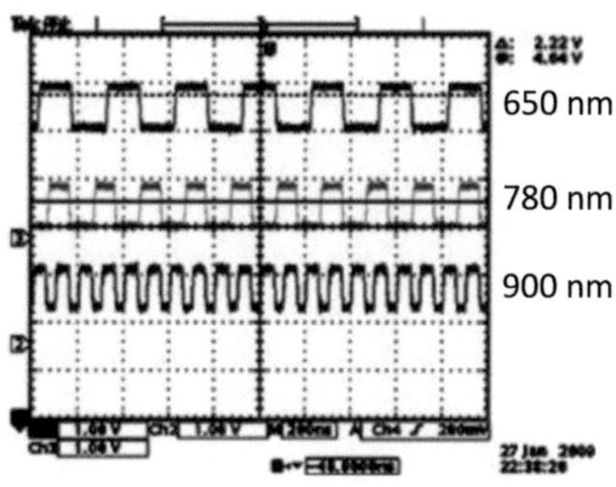


Fig. 17. Waveforms for three wavelength channels in WDM optical data transfer.

### 5.3 Beam steering

The prototype system of the beam steering optics with the fabricated beam steering lens was demonstrated and characterized experimentally. Figure 18 shows a simplified measurement setup. Output beam angle, radial and orthogonal spot size, and output power were measured.

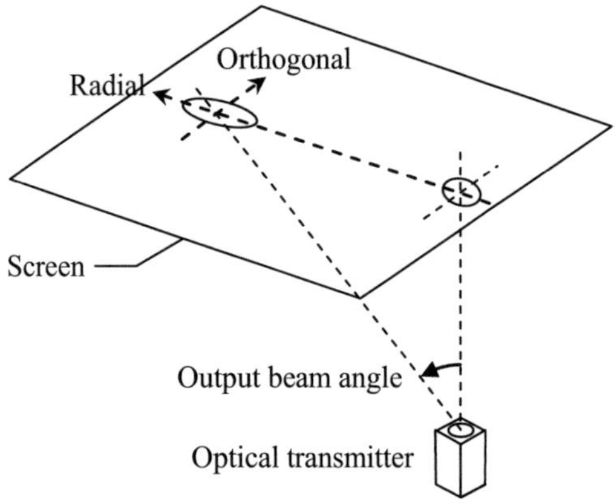


Fig. 18. Configuration for characterizing the beam steering lens.

Figure 19(a) shows a measured relationship between the input and the output beam angles of the beam steering lens. The maximum output beam angle was about 66 degrees in the experiments, which was large enough to cover a 5 m-by-5 m room when the vertical distance between the hub and a node was larger than 2 m. The maximum angle gain was 3.4, which showed a good agreement with the design.

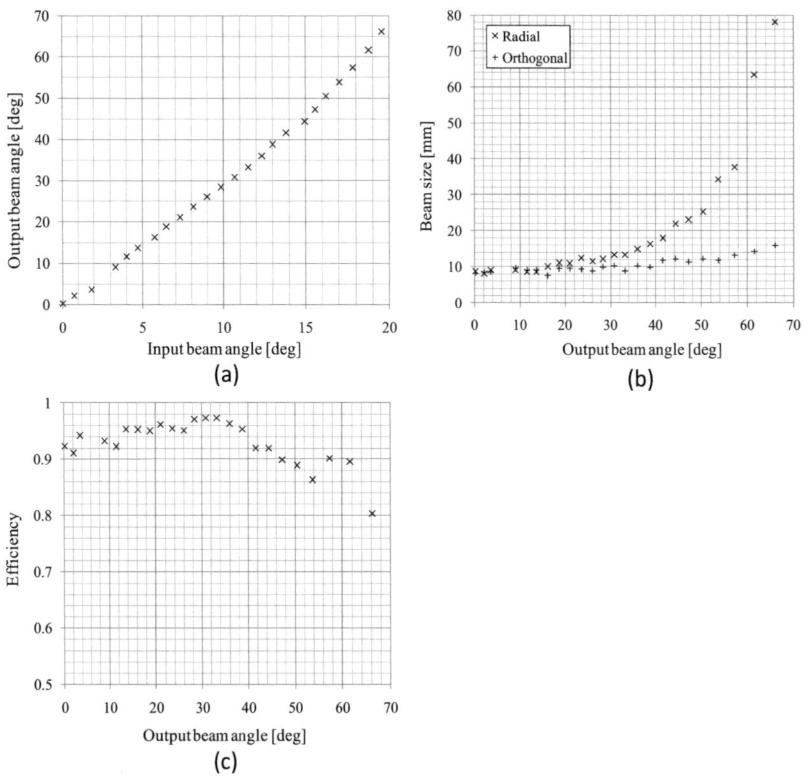


Fig. 19. Experimental results of the scan lens: (a) relationship between input and output beam angles, (b) beam sizes on the ceiling, and (c) power efficiency.

The beam size on the receiver plane is a significant concern because it defines received power, namely, the signal integrity. The beam size strongly depends on the radiation angle and the size of the laser source. Figure 19(b) shows the beam sizes measured on the ceiling 173 cm above the beam steering optics. The radial beam size increases as the output beam angle becomes large.

Figure 19(c) shows the optical power efficiencies of the beam steering lens. The minimum efficiency at the maximum output beam angle was bout 0.8 (-2dB). Note that the sidelobes of the beam were included in the measured power, and the actual efficiencies were possibly smaller than the results in Fig. 19(c).When we assume that the received optical power is in proportion to the beam area, the optical power variation at the receiver introduced by the beam steering optics is about 13dB, which can be tolerated by the dedicated CMOS imager by a gain control of the photoreceiver circuits. In conclusion, the experimental results showed that the prototype beam steering optics operated successfully.

## 6. Conclusion

A space- and wavelength-division-multiplexing (SDM and WDM) indoor optical wireless LAN system, which is based on a dedicated CMOS image sensor to realize a compact, high-speed, and intelligent nodes and hub, was described. The dedicated CMOS image sensor can detect multiple fast optical data concurrently as well as captures ordinary images from which positions of communication nodes or the hub is obtained. In this Chapter, with the CMOS image sensor, an application of WDM technique to downlinks was demonstrated. 64x64-pixel custom CMOS image sensor with 4-channel concurrent data acquisition function was fabricated. Experimental results showed that the CMOS sensor received 10Mbps x 3ch WDM data while capturing ordinary images. A wide-angle beam steering optics with a beam steering lens for amplifying the output beam angle was demonstrated. A prototype beam steering lens optimized for a near-infrared wavelength of 850 nm was fabricated. Experimental results showed that the maximum output beam angle was about ±60 degrees, which was enabled to cover a 5m-by-5m room (for the ceiling 2.0 m above the nodes), and the optical power efficiency was larger than 0.8. The received optical power variation caused by the power efficiency fluctuation and the beam distortion was roughly estimated to be 13dB, which was tolerated by the dedicated CMOS imager.

## 7. Acknowledgment

This research was promoted by Strategic Information and Communications R&D Promotion Program (SCOPE) by Ministry of Internal Affairs and Communications, and was partially supported by "Global COE (Centers of Excellence) Program" of the Ministry of Education, Culture, Sports, Science and Technology, Japan. This work was also supported by VLSI Design and Education Center(VDEC), the University of Tokyo in collaboration with Cadence Design Systems, Inc. The VLSI chips in this study were fabricated in the chip fabrication program of VLSI Design and Education Center (VDEC), the University of Tokyo in collaboration with Rohm Corporation and Toppan Printing Corporation.

I am grateful to Jun Ohta at Nara Institute of Science and Technology (NAIST) for valuable advice, and to Eiji Tanaka at Panasonic Electronic Devices Co., Ltd. for kindly supporting design and fabrication of the beam steering lens. I would like to heartily appreciate tremendous devotions of the students at NAIST who were involved in this study.